Industrialized Building System: The Malaysian
Approach

# Industrialized Building System: The Malaysian Approach

**Maryam Qays Oleiwi**

Filspay Academy
2015

First Printing: 2015

ISBN 978-1-312-78264-8

Filspay Academy (867118-D)

16-2 Jalan Putra kajang 3, Taman Putra kajang 3
 Kajang, 43000, Selangor, Malaysia

www.archicivi.com

## Dedication

Specially dedicated to

My kind parents, Qays and Samyah
My faithful husband Saif,
My elder son Yousif
My lovely daughter Teeba
And my little cute baby Haneen

For your everlasting love and care....

# Table of Contents

# Acknowledgements

In the name of Allah, the most gracious, the most merciful

Firstly, all praises and gratefulness are to Almighty "ALLAH" for granting me the chance, the will, the strength and the endurance to accomplish this book.

Secondly, I would like to express my sincere gratitude to my husband Dr. Saif Yousif Abdullah for his invaluable support, understanding, patience and kindness.

My utmost gratitude is to the honorable Prof. Ir. Dr. Kamal Nasharuddin Mustapha whose encouragement and guidance enabled me to develop an understanding of the subject and go through my study.

Immense thanks to Assoc. Prof. Dr. Tan Juat Hong for the help and co-operation that he showed. Also, I would like to thank Associate Prof. Dr. Hashem Almattarneh and Dr. Yousif Al-Sewaidi for their valuable suggestions and comments.

Great thanks to my parents and siblings for their continuous prayers, love, help and encouragement.

Moreover, I would like to express my gratitude to Filspay Academy and their programmers who helped me in programming IBS online survey system which is used to collect the data needed in my research.

Lastly, thanks to all the friends who helped me during working on this book.

# Preface

Industrialized building system (IBS) was proposed in several studies as a solution to meet the intensive demands and requirements of buildings. The objectives of this book are to explain the current status of the industrialized building system in Malaysia, and also to present general conclusions along with the vision and suggestions of several experts towards future developments.

The contents of this book are based on my Mater Thesis about industrialized building system in Malaysia, which was carried out in the Civil Engineering Department of Universiti Tenaga National (UNITEN), Malaysia.

This book is divided into three main parts. The first part includes the definitions of industrialized building system, the different types of industrialized building system and other general information about this system. The second part includes the constraints that faced the application of industrialized building system in Malaysia, the recent studies and researches about this system in Malaysia, and the well-known projects that have been executed using this system in Malaysia. The third part includes suggestions about improving the IBS in Malaysia.

Maryam Qays Oleiwi

# Introduction

The simplest form of the industrialized building system (IBS) was found in the United States of America, where steel beams were combined with pre-cast slab panels and used in rapid construction of skyscrapers during the early part of the previous century. In Europe, after the devastation that resulted after the Second World War, several European countries adopted prefabricated systems due to the intensive shortage materials and human resources.

In Malaysia, the Traditional Malay House (TMH) was a very common type of housing in the past. The components of these houses were prepared before construction, including the columns, beams, walls, floors and roofs. These houses provided the concept of modularity and repetitiveness that had helped in the development of construction industrialization.

Traditional Malay House (TMH)

The first manufacturer of standard designs of traditional Malay houses in Malaysia was Kayu Sedia (KS) company, which was established in 1968 (Ismail, 2007). One of the KS products was the readymade houses with affordable prices and variations in design. Kayu Sedia Company was considered an IBS manufacturer because of standard design modules, applying standard dimensioning systems (Imperial foot and inch), mass-production based on standard design modules, and in factory and high-quality production (Ismail, 2007).

The prosperity and tremendous increase in the Malaysian population after the independence had resulted in increasing demand for housing. The conventional construction method was not enough to meet these demands due to the slow pace of construction and high costs (Agus, 1997). Therefore, the Malaysian government took great efforts to adopt the IBS. The start was in the sixties of the last century, when the minister of housing and local government traveled to many European countries to utilize from their experience in the housing field. Subsequently, the government adopted two pilot projects. The first project was in Kuala Lumpur and the second was in Penang, both using precast concrete elements to build these high-rise low-cost flats (Din, 1994).

Industrialized building system (IBS) was proposed in several studies as the solution to meet the intensive demands for buildings. Several studies mentioned that using this system can provide valuable advantages such as providing in-factory quality control, saving construction time, minimizing the dependency on foreign workers, increasing safety during work and increasing cleanness and neatness at the construction site. These advantages have encouraged many of Malaysia's World-class developers to use industrialized building systems in their projects. Furthermore, the encouragements and the great efforts that were introduced by Construction Industry Development Board (CIDB) Malaysia played a pivotal role in raising the application of IBS in Malaysia (CIDB, 2003b).

# Chapter 1: Introduction on Industrialized Building System

## 1.1 Definitions of industrialized building system

A number of studies have written in the definition of industrialized building system. According to (CIDB, 2003a), industrialized building systems can be defined as a construction system through which components are manufactured in a factory, on or off site, positioned and assembled into structures with minimal additional site work. Rahman and Omar (2006) mentioned in their study that industrialized building system (IBS) is a construction system that is built using pre-fabricated components. The manufacturing of the components is systematically done using machines, formworks and other forms of mechanical equipment. The components are manufactured offsite and once completed is delivered to construction sites for assembly and erection.

On the other hand, industrialization process in construction can be defined as mass producing all building components such as walls, floor slabs, beams, columns and staircases either in factory or at site under strict quality control and minimal on-site activities (Trikha, 1999). Warszawski (1999) have defined the industrialization process as an investment in equipment, facilities, and technology with the purpose of increasing output, saving manual labor and improving quality. While Sarja (2005) mentioned that the industrialization of buildings means the application of modern systematized methods of design, production planning and control as well as mechanized and automated manufacturing processes.

All these definitions reflect the same concept of manufacturing the components before construction (ether in factories or at site), then, the components are installed at site using high machinery techniques.

## 1.2 Types of industrialized building systems

According to CIDB (2003b), from the structural classification point of view there are five main types of industrialized building system used in Malaysia. The first type is a precast concrete framing, panel and box system which includes precast columns, beams, slabs, 3-D components (balconies, staircases, toilets, lift chambers), permanent concrete formwork, etc.). Figure 1.1 shows the use of precast concrete system in building.

Figure 1.1: Precast concrete system

The second type is a steel formwork system which contains tunnel forms, beams and columns moulding forms, permanent steel formworks, metal decks, etc. Figure 1.2 shows a steel formwork system.

Figure 1.2: Steel formwork system

The third type is a steel framing system which includes steel beams and columns, portal frames, roof trusses, etc. Figure 1.3 shows steel framing system.

Figure 1.3: Steel framing system

The fourth type is a prefabricated timber framing system which includes timber frames, roof trusses, etc. Figure 1.4 shows timber framing system.

Figure 1.4: Prefabricated timber framing system

The last type is a block work system which includes interlocking concrete masonry units (CMU), lightweight concrete blocks, etc. Figure 1.5 shows block work system.

Figure 1.5: Block work system

### 1.3 The advantages of industrialized building system

A number of researchers clarified the advantages of using industrialized building system in their studies. These advantages can be listed as below:

### 1.3.1 High quality

Industrialized building systems can provide high quality-controlled products due to employing skilled and semi-skilled workers with specific scope of works to improve efficiencies and reduce errors (Bahri, 2009). This system is also suitable for countries like Malaysia where the rain extends over several months every year. In this system, the casting process is unaffected by weather elements due to controlled environment either in the factories or covered casting yards at sites (Thanoon et al., 2003). Furthermore, using high mechanized technology and material selection can improve the quality of the products (Warszawski, 1999). The components that do not meet the specifications can be rejected before delivering it to a construction site.

### 1.3.2 Cost reduction

Using an industrialized building system can minimize the total cost of the projects due to the ability of reducing the number of on-site workers significantly reducing labor cost for contractors (Masod, 2005). Moreover, the ability to use the moulds which are made of steel, aluminum, etc. for several times contribute to minimize the total cost (Bing et al., 2001). In addition, the exemption of the construction levy imposed by CIDB can be given to housing developers in the private sector who use IBS components over 70% (Bahri, 2009).

Furthermore, rectification works can be minimized because of closely checking and controlling in-factory environment (CIDB, 2005a). In addition, there is an ability to reduce the waste materials and minimize the cost of transferring waste materials duo quality control (Bahri, 2009). Also, IBS waste can be recycled and reused to produce alternative aggregates (Kamar and Hamid, 2008).

### 1.3.3 Time reduction

It was proved that using industrialized building system can provide faster completion of projects due to advanced mechanized production and simplified installation process (Masod, 2005). Furthermore, by using this system, the production of components in factories can start while the construction site is under earthworks (CIDB, 2005a). Moreover, the speed of manufacturing IBS components is unimpeded by adverse weather conditions.

### 1.3.4 Increasing safety

Safety in construction is a prominent matter during pre-construction, construction and post construction (Ghani et al., 2007). IBS can promote safe and systematic working environment due to minimizing the number of workers, materials and waste at the site (CIDB, 2005a).

### 1.3.5 Increasing environmental aspects

The use of industrialized building systems in projects can provide cleaner sites due to systematic components storage and timely material delivery (CIDB, 2005a). Furthermore, there will be a reduction of construction materials at site and its' waste duo to casting in factory (Masod, 2005). Moreover, the use of timber formworks and props at the site can be minimized due to casting in factory (CIDB, 2005a). The dust and suspended particles can be decreased due to in factory prefabrication, thus reducing air pollution at the construction site. Noise at the site can also be minimized as there is no need for erecting scaffoldings and formworks and dismantling them later (Trikha &Ali, 2004).

### 1.3.6 Social advantages

By using industrialized building systems, the manual labor on construction site can be reduced (about 40 - 50% compared to conventional method) especially when high degree of mechanization is involved (Warszawski, 1999). This can reduce the dependency on unskilled foreign workers, reduce money remittances to their countries and reduce their social problems, low quality works, delays, and diseases (CIDB, 2009a).

### 1.4 Principles and features of industrialized building system

Many earlier scholars have written in details about the principles of industrialized building system, however, the main ones is listed by Masod (2005). We can some up these principles as follows:

- Prefabrication: using prefabricated and precast components,

-Industrialization: factory production of components,

- Standardization: using standardized dimensions,

- Repeatability: design using repeated elements and

- Modularization: design using modular coordination concept.

Recently, Warszawski (1999) listed the main features of industrialized building systems as:

- Production process of components in factory which can provide better specialization and organization due to controlled environment.
- Design, production, and erection onsite must be viewed as parts of an integrated process which has to be planned and coordinated accordingly.
- Incorporation of building works into large prefabricated assemblies with minimum erection, jointing and finishing work onsite.
- Equipment used onsite is highly mechanized; in concrete work, large standard steel forms, ready-mixed concrete and concrete pumps.

### 1.5 The sequence of construction for industrialized building system

Industrialized building system is considered as an innovative approach among the construction methods commonly used. It offers an alternative to the existing conventional building system. As compared with conventional wet construction method, industrialized building system has its benefits of shorter construction time, saving in labour, saving in material, better quality control and the immunity to weather changes. The integration between the designers, planners, manufacturers, transporters and erection engineers at site is an important issue when using the industrialized building system.

The first stage of any IBS project is the design stage that is carried out where the IBS components are designed according to the specifications of the project. Next, the components are manufactured either in factories or at construction sites with high machinery techniques. When the IBS components are manufactured in factories, they have to be transported from the factories to construction site for assembling process. IBS components have to be delivered at construction site within the exact time to avoid delay in construction process.

Figure 1.6: Moulds preparation at an IBS factory

Figure 1.7: Casting process at an IBS factory

Figure 1.8: IBS components are ready to delivery

Figure 1.9: Delivering IBS components to the construction site

When the IBS components are manufactured at construction site, it is preferable that the casting yard to be covered to avoid weather effects. It is necessary at casting yard to have machineries for automated concrete production, reinforcement cutting and bending, reinforcement placement in moulds, concreting, cutting and sawing to the correct length of the elements if re- quired.

Figure 1.10: Reinforcement cutting and bending at construction site

Figure 1.11: Concreting process into IBS steel mould at construction site

Figure 1.12: Leaving IBS components to dry

Figure 1.13: Removing the reusable steel moulds that used to produce IBS components

Figure 1.14: Lifting of IBS Concrete components

Figure 1.15: Stocking of IBS Concrete Column

One of the most important characteristics of IBS method is that IBS components are prefabricated before construction. According to Chew and Michael (2001), prefabrication process means breaking a whole housing unit into different components such as the floors, walls, columns, beams, roofs, etc. and having these components separately prefabricated or manufactured in modules or standard dimensions in a factory or on site. The issue of high quality and aesthetical aspects can be controlled during the processes of controlled fabrication when the components are produced.

At site, IBS components are assembled accordingly with the assistance of cranes. The machineries that are used in this stage have been developed to lift precast elements and align them in their exact position at the construction site. However, using cranes may limit the maximum size and dimensions of the prefabricated components to make them easier to deal with.

Figure 1.16: Erecting IBS components into their positions

(Trikha & Ali 2004) mentioned that the integration of erection and assembly process with the manufacturing and production as well as the design and information process into a fully integrated system through a centralized control system are possible in the future of this industry, if it were to progress beyond industrialization to computer integrated construction. The reduction of construction waste with the usage of the standardized components and less in-site works provides cleaner site due to lesser construction waste (Basri 2008).

According to Chew and Michael (2001), there are two types of prefabricated systems in the market, namely; fully prefabricated system and partially prefabricated system.

A fully prefabricated system refers to the components produced in the factory or at site, and later positioned into their right place for erection. A fully prefabricated system consists of three categories, namely; panel system, frame system and box system.

Figure 1.17: Fully prefabricated system

Partially prefabricated system is another type of construction system where certain element that can be standardized are prefabricated in the factory, whereas other components are cast in-situ using conventional method. In this construction method, certain elements as such wall panels, slab panels and staircase are considered as precast components, while the other elements like columns, beams and the foundations are conventionally produced due to ease and speed of construction. According to Chew and Michael (2001), this system usually gives more rigid construction and better water tightness characteristic, which is considered as a great problem with the usage of panel system and frame systems.

Figure 1.18: Partially prefabricated system

## 1.6 Comparison between industrialized building system and the conventional method

Many organized bodies perceived that industrialized building systems may reduce building costs; however, several studies proved that this system is more expensive than conventional methods. A survey carried out by Kadir et al. (2006) included a sample of residential projects reveled that there was a significant difference between industrialized building system and the conventional method in terms of actual labor productivity, crew size, and cycle time while the structural cost between conventional method and industrialized building system was found to be insignificantly different. This clarified why large number of contractors preferred to use the cheaper conventional method. The authors suggested that the government have to impose a legislative requirement on the use of industrialized building system or re-define the market so as to earmark a set quota of IBS projects.

Haron et al. (2005a) proved in their study that fully prefabricated building systems were more expensive than conventional method. The study demonstrated that the conventional method was more economical about RM 4.72 per gross floor area ($ft^2$) for single story house and about RM 4.71 for double story house. Figure 1.6 shows the comparison between cost per gross floor area ($ft^2$) of conventional and fully prefabricated system.

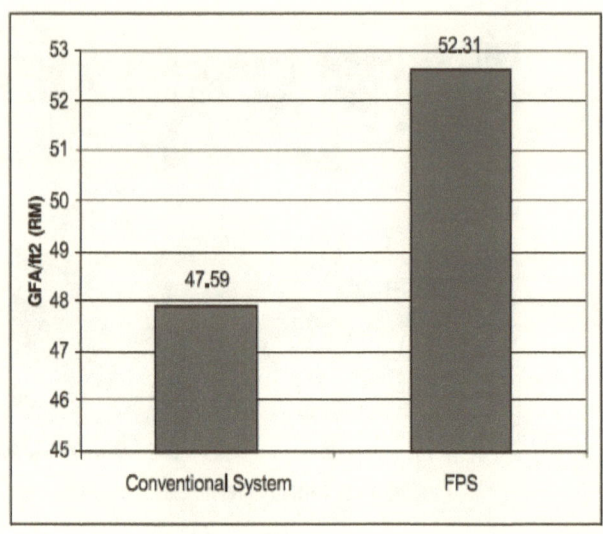

Figure 1.19: Comparison between cost of gross floor area per (ft²) of convention and
fully prefabricated system (Haron et al., 2005a)

Another study showed that the formwork system, which is considered a
one type of industrialized building system, was more expensive as compared
to the conventional method. The main reasons were that the conventional
method provided better negotiation chances to obtain the most competitive
tender price appropriate to a developer's budget. Moreover, the conventional
approach could confer more flexibility to choose alternative construction
materials at lower prices (Haron et al., 2005b). Another comparative study
carried out to compare between the composite construction method (IBS and
conventional) and conventional method. The study demonstrated that the use
of a conventional method in executing one unit single story for low-cost house
was cheaper than using a composite construction system worth about RM
52.60 per gross floor area ($m^2$). In spite of that, the composite system offers
better quality structure and finishing and shorter construction time to
complete the projects (Haron et al., 2005c).

However, it was proven that the interlocking precast concrete block
system was cheaper, or at least comparable, to conventional construction in

Malaysia. The use of this system is increasing in popularity, especially for low-rise dwelling construction (CIDB, 2008).

In terms of labor productivity, a study carried out by Kadir et al. (2005b) illustrated that the labor productivity of precast concrete system was the highest among other systems, followed by cast in-situ half tunnel system, cast in-situ table form system and the conventional building system was the lowest labor productivity. In terms of crew size of workers, the conventional system required 7% more than the industrialized building system. Furthermore, the conventional building system required 53% more cycle time than of precast concrete system.

Another study carried out by Mian (2006) found that precast concrete system was capable of saving 29% of the time required for the construction compared to that of conventional systems. In terms of labor productivity, the conventional system was laborious and needed large number of workers to complete the construction cycles for instance formwork fabrication, reinforcement bar or steel cage fabrication, formwork installation, reinforcement bar installation, concrete placement, and formwork dismantle etc. (Mian, 2006). The application of industrialized building systems does not require much wet trades at the site; therefore, the total number of workers could be reduced. However, many contractors preferred using the conventional construction system because of the low costs of foreign workers in Malaysia (CIDB, 2007).

Although many members of the local authorities were keen on the idea of using industrialized building systems in the construction sector, the majority of the industry stakeholders were hindering this system due to their resistance to change to a new system and insufficient information to encourage changing (Hamid et al., 2008).

A research conducted by Basri (2008) found that the productivity of precast construction method was affected by overtime working hour, precast components delivery practices, maximum temperature of the day, hours of raining of the day, crew size and repetitive activity. When the productivity was affected, the total time needed to complete the project could be affected, and therefore, delay could be occurring. Badir et al. (2002) mentioned that the main reasons of delay in completing projects in Malaysia are supply delay, bad weather, lack of experience and shortage in raw materials.

The main reasons of preserving the use of conventional methods in the Malaysian construction sector is because of the intensive capital cost of industrialized building systems compared to the capital cost the conventional labour. Furthermore, the failure of early IBS projects introduced in Malaysia is due to using western systems, which were not suitable for Malaysian climates and did not consider the aspect of serviceability such as the need for wet toilets and bathrooms have brought bad reputation to this system (Rahman and Omar, 2006). Moreover, shoddy installations due to lack of workers' experiences to install IBS components during that time, and the improper maintenance of the low-cost housing have caused problems of leakage. All these reasons have further contributed to the poor image and negative impression of industrialized building systems in Malaysia (CIDB, 2005a).

In case of Hong Kong, a study carried out by Jaillona and Poon (2008) included a questionnaire survey that was exposed to experienced professionals and case studies of seven recent residential and non-residential buildings were conducted. The findings revealed that the use of prefabricated concrete systems could achieve significant economic, social and environmental benefits, especially in a dense urban environment like Hong Kong. Another study carried out by Jaillona and Poon (2009) demonstrated that prefabricated systems provided significant advantages compared with conventional systems in Hong Kong, such as improved quality control, reducing construction time by about 20%, reducing construction waste by about 56%, reducing dust and noise on-site, and reducing labor requirement on-site by about 9.5%.

# Chapter 2: The Experience of Malaysia and Other Countries in Industrialized Building System

The technology transfer adopted from Europe, Japan and Singapore had accelerated Malaysian efforts towards the implementation process of industrialized building systems (Hamid et al. 2008). The experience of several developed countries and Malaysia in the area of industrialized building system can be listed as below:

## 2.1 Industrialized building system in the United States of America

In the United States of America, prefabricated systems gained wide-spread market share with 30% use in the housing sector during the nineties of the 20th century. Most low-rise housing projects used concrete precast systems and timber framing systems intensively, especially in the area of environmental hazards such as hurricanes and tornados. Although prefabricated systems were encountered by some hindrance such as the shortage in risk of water penetration, difficulties of installing insulation and plain appearance of panels, however, the improvement of advanced technology introduced solution such as improved moulds, availability of rigid foams, improved concrete mixing techniques and an improved surface finishing (Glass, 1999).

## 2.2 Industrialized building system in Japan

The application of industrialized building systems started in Japan since the sixties of the 20th century. However, the prefabricated houses consisted of 20% of the total houses built from April 1999 to March 2000. The most dominated system in prefabricated markets was the steel framing system, which represented 73% share, followed by the wood framing system which represented 18%, while the reinforced concrete framing represented only 9%. The growth of steel framing housing was 3% and wood framing housing was 2%, while the concrete framing housing withdrawing was 12% (Nagahama, 2000). Table 2.1shows this information.

Table 2.1: The prefabricated housing market share in Japanese fiscal year 1999 (April 1999 - March 2000) (Nagahama, 2000)

| No. | Framing structure | Prefabricated Market Share, % | Growth Rate, % |
|-----|-------------------|-------------------------------|----------------|
| 1   | Steel framing     | 73                            | +3             |
| 2   | Wood framing      | 18                            | +2             |
| 3   | Concrete framing  | 9                             | -12            |

## 2.3 Industrialized building system in Britain

After the Second World War's devastation, Britain suffered from critical deficiency of housing units, skilled workers and construction materials. The need for using a fast and economic construction system was very urgent. Therefore, the use of prefabricated system was adopted and turned to be more popular to solve housing problems. According to Glass (1999), over 165,000 prefabricated dwellings were built ranging from small single bungalow to large high-rise buildings by 1960. In 1999, the prefabricated concrete represented 25% of the construction industry products in the British market. The production included several kinds of precast systems such as structural blocks, paving, cast stone, suspended floor and architectural cladding. The suspended floors production represents the higher use annually. Currently, prefabricated systems play a vital role in the British construction industry. Glass (1999) indicated that there was no market resistance to the idea of using this system in the housing field. The researcher expected that prefabricated systems would monopolize the construction markets and overwhelm the conventional method in the near future.

## 2.4 Industrialized building system in Denmark

After the severe devastation of the Second World War, Denmark suffered from colossal privation in materials and human resources. The alternative choice to solve that problem was to use the prefabricated construction system in the construction sector. In 1958, Denmark started precast modular housing execution. Denmark achieved great implementations in prefabricated system field since its starting (Kadir et al., 2005a). According to Gibbons (1986),

80% of the detached houses produced till the mid-1960 used this system. The highest production was occupied by the panelized system. Jespersen & Son and Larsen & Nielsen were examples of Danish international contractors who executed many large-scale projects all over the world using prefabricated components which were produced in the local factories in Denmark.

## 2.5 Industrialized building system in Singapore

In the 1960s, there was an urgent need to execute a great number of houses in Singapore. Therefore, The Housing and Development Board (HDB) adopted the prefabrication concept. In 1963, the HDB launched the first pre-fabricated project consisted of 10 blocks of standard 16 story flats using French large panels. Unfortunately, the project had to be complete by using conventional methods due to management and technical problems (Tat and Hao, 1999). In spite of that failure, HDB did not despair and improved its way towards enhancing the experience of prefabricated systems. In the seventies, 100,000 dwelling units were needed to be achieved. Therefore, the HDB took brave paces by executing these dwelling units using prefabricated system ap-plications. As a result of the hard work, the HDB achieved good experience in public housing programs using prefabricated method and gained substantial market in the construction industry during the nineties (Tat and Hao, 1999). Figure 2.1 shows HDB precast concrete implementation.

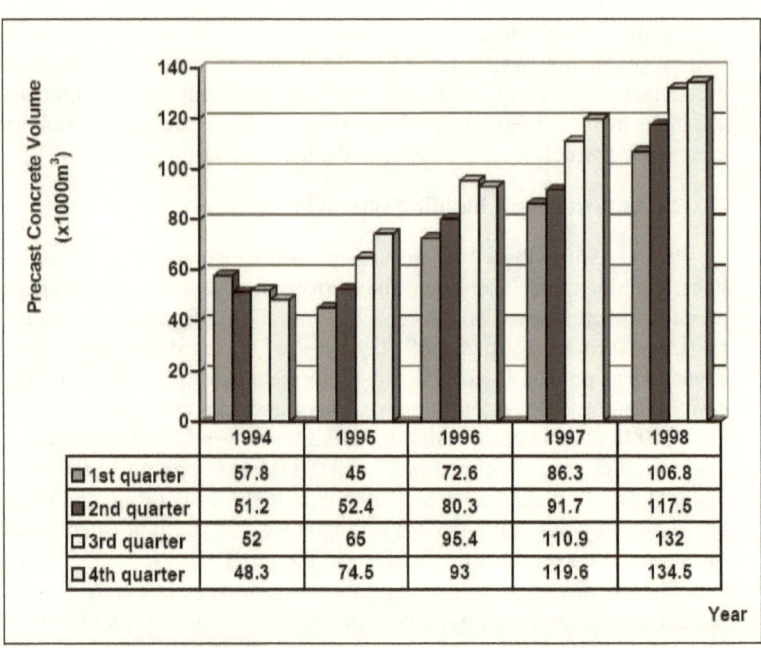

Figure 2.1: HDB precast concrete implementation (Tat and Hao, 1999)

The main achievement of HDB in prefabricated area was the establishment of the Prefabrication Technology Centre (PTC) in 1994. The principal activities of PTC were to conduct research and development of advanced and innovative construction materials and systems, design and produce prefabricated building products, manage and supply prefabricated building products, conduct training and license its intellectual property rights (Mian, 2006). The HDB added significant innovations in the area of prefabricated systems, which include architectural precast facades, precast pre-stressed composite floor systems, prefabricated bathroom systems, ferro-cement cladding systems and pre-cut and pre-bend reinforcement bars (Thanoon et al., 2003).

### 2.6 Industrialized building system in Thailand

The prefabricated system in Thailand witnessed rapid development since it started in the seventies of the 20th century, and the application of this system - as an innovative construction method in Thailand - gained high popularity (Sharma 2004). In the middle of 2004, the National Housing Authority

(NHA) of Thailand approved the usage of the PLP (Precast Large Panel) Construction. This system consisted of reinforced precast concrete panels, precast slabs, foundations, beams and columns in very few locations.

PLP system provided several advantages to the construction sector in Thailand. Compared to the conventional methods, the total construction time reduced when using PLP. According to Sharma (2004), the time required for casting, lifting, erection and completion of structural system for a typical 2-3 bedroom two storey houses is about 2 to 4 days. This application was good evidence of the fastness of this system compared to the conventional method. Figure 2.2 shows the application of PLP in the construction industry in Thailand.

PLP components layout on bed

Casting of PLP components

Stacking of PLP components

PLP components are lifted into position

An almost done PLP house

The magnificent product of PLP

Figure 2.2: The application of PLP in the construction industry in Thailand (Construction Photos (Sharma, 2004)

In term of cost, PLP provided less or comparable structural cost with low initial investment. Furthermore, the quality of buildings was increased when using PLP system due to employing skilled workers. Therefore, this system provided buildings durability and long-term performance. PLP also provided automation and modularization (Sharma, 2004)

## 2.7 Industrialized building system in Malaysia

Industrialized building system was introduced in Malaysia over 40 years ago. After the successful visit of the Minister of Housing and Local Government to several European countries in 1964, the government initiated the first IBS pilot project in Jalan Pekeliling in Kuala Lumpur consisted of 3000 units of low-cost flats and 40 shop lots using the large panel Danish System (Din, 1994). In 1965, the second IBS pilot project was implemented in Jalan Rifle Range in Penang and consisted of 3,699 units and 66 shop lots using French Estiot System (Din, 1994).

The design of the European systems was inefficient to meet the need of Malaysian style of life such as wet toilets and bathrooms. Therefore, the problems of leakage and sealant contributed to bringing a bad reputation to industrialized building systems in Malaysia (Rahman and Omar, 2006). Moreover, shoddy installations contributed to these problems. Today, precast technology has since improved and these problems became a part of the past decade (CIDB, 2005a).

A comparison between the industrialized building system that used in these two pilot projects and the conventional method that was used in other projects that was constructed in that time was done by Din (1994) to evaluate the advantages and disadvantages of industrialized building system and whether this system was more effective than conventional method in terms of quality control, cost, speed of construction, and labor requirements.

In terms of quality, industrialized building systems (IBSs) appeared higher quality when compared with the low-cost buildings constructed using conventional method. The finishing of IBS interior walls was more aesthetical than conventional walls constructed using cement-sand hollow blocks. Concerning cost, the first pilot project which was located in Kuala Lumpur was more expensive than conventional systems for about 8.1%. However, the cost of the second pilot project was cheaper by approximately 2.6% compared to conventional system housing project completed around that time.

Furthermore, industrialized building systems saved more time when compared with conventional methods. The construction time of the two pilot projects took 27 months including establishing the plants. It was found that the use of industrialized building systems, in terms of time saving, needed shorter constructions time and the prefabrication process was unaffected by weather conditions. In these two pilot projects, industrialized building system saved labor in about 30% to 40% compared to conventional practice particularly in the field of brick layers, plastering and carpentry (Din, 1994).

After that, several attempts have been introduced in the market to avoid the mistakes that occurred in the two pilot projects. Wet joint technology was identified to be more suitable to be used in Malaysian tropical climates and it was also better to use for the bathroom types, which were relatively wetter than those in Europe (Rahman and Omar, 2006).

In 1978, the Penang State Government launched another 1200 units of housing using prefabrication technology. Two years later, the Ministry of Defense adopted large prefabricated panel construction system to build 2800 units of living quarters at Lumut Naval Base (Trikha and Ali, 2004).

In the early eighties up to the nineties, the use of structural steel components had a new role, especially in high rise buildings in Kuala Lumpur. The usage of steel structure gained much attention with the construction of 36- storey Dayabumi complex that was completed in 1984 by Takenaka Corporation of Japan (Rahman and Omar, 2006).

Figure 2.3: Dayabumi complex

During that period of time, Perbadanan Kemajuan Negeri Selangor (PKNS) - a state government development agency- acquired precast concrete technology from Praton Haus International based on Germany to execute low cost houses and high cost bungalows for the new townships in Selangor (Hashim, 1998).

The nineties witnessed several successful projects after the development of precast steel and hybrid construction, such as the construction of the Kuala Lumpur Convention Centre (steel framing and trusses with precast slab), KL Sentral (steel structure & precast hollow core), Bukit Jalil Sports Complex and Games Village, the LRT lines and tunnels (steel structure and precast hollow core) and Kuala Lumpur International Airport (steel roof structure and formwork system for slab) (CIDB, 2003a).

Figure 2.4: Bukit Jalil Sports Complex and Games Village

However, the development of industrialized building system was in slow pace. CIDB survey 2003 revealed that the usage of industrialized building systems in the local construction industry was only 15% (CIDB, 2003a). IBS Mid Term Review 2007 illustrated that the completed project using industrialized building system in the year 2006 was only 10% (Hamid et al., 2008) which is considered as a very small percent as compare to IBS Roadmap forecasting IBS projects of 30% in 2004 and 70% in 2008 (CIDB, 2003b).

Moreover, Malaysian owned production of industrialized building system was very small compared to the production that originated from other countries. It was reported that Malaysian production of industrialized building systems was 12%, whereas a greater portion was originated from the United States, Germany and Australia with market share of 25%, 17% and 17% respectively. This result indicates that there were considerable needs to improve the production of industrialized building systems in Malaysia (Badir et al., 2002). Figure 2.5 shows the sources of industrialized building system in Malaysia according to the origin of countries.

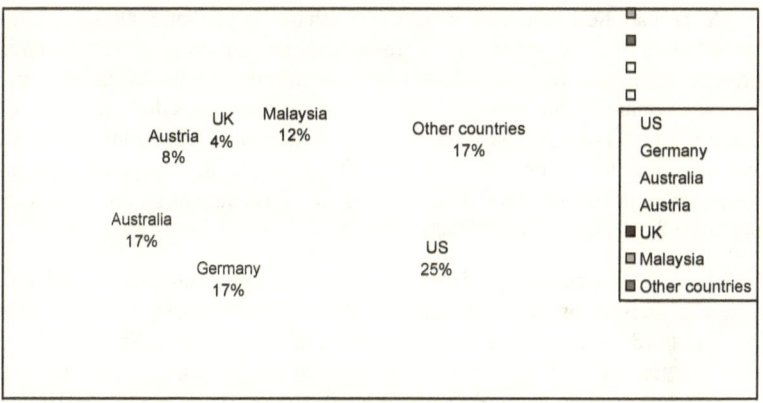

Figure 2.5: The sources of IBS in Malaysia according to the origin of countries (Badir et al., 2002)

The Malaysian government has been highly committed the implementation of industrialized building system. In 2003, the Cabinet of Ministers endorsed the IBS Roadmap 2003-2010 to be the blueprint document to achieve total industrialization of Malaysian construction sector.

The first strong indication by the government after the endorsement on IBS Roadmap was through the 2005 Budget announcement back in September 2004. It was announced that all new government building projects were required to have at least 50% IBS content, which was calculated through the IBS Score Manual developed by CIDB (Badawi, 2004). The reason of making the use of industrialized building system compulsory for government's pro- jects was to create sufficient momentum for the demand of IBS components (Shaari, 2006).

The second announcement was on the levy exemption for housing projects that have at least 50% IBS components to attract the developers in private sector (Badawi, 2004). During 2006 Budget announcement, the government offered tax incentive through Accelerated Capital Allowance (ACA) for IBS manufacturers to purchase moulds used for the production of precast concrete components. In addition, it was mentioned in 2006 Budget that the government would ensure that all IBS components used in public projects were complied with Malaysian Standard MS 1064 to encourage and facilitate standardization of IBS components (Badawi, 2005).

After that, the announcement of the Ninth Malaysia Plan 2006 - 2010 has strengthened the commitment of industrialized building system agenda through emphasis on of industrialized building system in public and affordable housing projects and offering more incentives for the users of standardized drawings based on the use of industrialized building system and modular co- ordination (Shaari, 2006). In 2008, the percentage of compulsory using of IBS components in government's projects was increased from 50 % to 70% (Bahri, 2009).

However, a statistical study conducted in Northern Malaysia (Kedah and Perlis) amongst housing developers revealed that the majority of respondents have not used industrialized building system in their projects. Statistics have revealed that only 31.6% of developers have used industrialized building system while the remaining was still using the conventional method (Nawi et al., 2007).

The success of industrialized building system in several developed countries was implemented by applying two policies: persuading home buyers about long-term energy saving, indoor air quality, and other health and com- fort related issues, and commitment of houses developers toward new innovated prefabricated technology. Therefore, Malaysian construction industry can mimic their steps and achieve great success in construction field (Thanoon et al., 2003).

### 2.7.1 IBS Roadmap 2003-2010

A survey conduction by CIDB (2003a) reveled that in spite of all the great efforts of Malaysian government to facilitate the use of industrialized building system in construction sector; the usage level of this innovative system in local construction industry was still low. In consequence, CIDB redesigned its strategies with the IBS Steering Committee's guidance to pro- pose the IBS Roadmap which was endorsed by the Cabinet of Ministers as a blueprint document. IBS Roadmap was based on the 5-M strategies which are Manpower, Materials (Components and

Machines), Management (Processes and Methods), Monetary (Economic & Financial) and Marketing    (CIDB, 2003b).

The roadmap aimed to increase the use of industrialized building system and execute the total industrialization in construction sector in Malaysia by the year 2010 (CIDB, 2003b). As mentioned in (CIDB, 2003b), IBS Roadmap introduced several suggestions to improve industrialized building system and modular coordination in Malaysia. Some of these suggestions can be mentioned as follow:

1. Providing the diploma and degree level students in higher learning institutions with industrialized building system and modular coordination courses.

2. Concentrating on IBS / MC technical training courses for architects, engineers, manufacturers, site supervisors and workers.

3. Minimizing the number of foreign workers gradually (from 75% in 2003 to 55% in 2005 and 15% in 2009).

4. Enforcing the use of modular coordination based on Malaysian Standard MS 1064 through Uniformed Building By Laws (UBBL) in 2004 by the local authorities.

5. Developing IBS publications and writings especially the Malaysian Standards (MS) and Construction Industry Standards (CIS).

6. Generate the IBS Verification Program and Resource Centre for selection process and certification, to monitor all IBS technologies offered for use by the Malaysian construction industry.

7. Persisting on Research and Development (R&D) efforts to produce local IBS innovations.

8. Enforcing utilization of IBS in government building projects - from 30% gradually increasing to 50% in 2006 and 70% in 2008.

9. Offering CIDB levy exemption for housing developers who use IBS components in their buildings.

10. Offering tax reduction incentives for the Bumiputera component manufacturers starting in 2005.

11. Conduct studies every 5 years to forecast the IBS components' needs for the Malaysian construction industry.

Unfortunately, the IBS Roadmap was launched at the end of the Eight Malaysia Plan. Therefore, most government allocations for development were utilized (Shaari, 2006). Only 54 out of 109 IBS Roadmap milestones were achieved until year 2007 (Hamid et al., 2008). This means that the percentage of achieving IBS Roadmap milestones until year 2007 was 49.5% of the total milestones.

### 2.7.2 IBS Roadmap 2011-2015

To continue the acceleration of industrialized building system in Malaysia, CIDB had replaced the former roadmap by the new IBS Roadmap 2011-2015. The policy objective was to impose high level intended outcomes of implementing IBS. The new roadmap focused on private sector adoption of industrialized building system through four policy objectives; which are quality, efficiency, competency and sustainability. A sustainable IBS industry will contribute to the competitiveness of the construction industry. The pillars of the new roadmap are as follows:

a) Good quality designs, components and buildings are the desired outcomes of IBS. Aesthetics should be promoted through innovations.

b) To ensure that, by using IBS, completion time of a building is speedier, more predictable and well managed.

c) To have a ready pool of component IBS professionals and workers throughout the entire project lifecycle: from design, manufacture, build and maintenance.

d) To create a financially sustainable IBS industry that balances user's affordability and manufacturers viability.

The goals for the Roadmap are encapsulated below:

1. To sustain the existing momentum of 70% IBS content for public sector building projects through to 2015

2. To increase the existing IBS content to 50% for private sector building projects by 2015.

It is hoped that the roadmap will drive the way forward for sustainable IBS adoptions; both in public and private sector.

# Chapter 3: Modular Coordination and Open Building System

## 3.1 Modular coordination

According to Trikha (1999), Modular coordination (MC) can be defined as a coordinated unified system for dimensioning spaces, components, fitting, etc., so that all elements fit together without cutting or extending even when the components and fittings are manufactured by different suppliers. Another study mentioned that modular coordination can be defined as a dimension and space coordination concept in which building, and components are placed at their designations based on the unit or basic module known as "1M" that equals to 100 mm (CIDB, 2009b).

The use of modular coordination is an important factor in IBS effective application, as it completes the industry through quality control and increase of productivity (CIDB, 2009b). In Malaysia, the rules of Malaysian Standard MS 1064 were set based on Nederland's Normalisatie Institute (NNI)'s NEN 6000 (Shaari, 2006). Malaysian Standard MS 1064 was planned to organize using practical approaches for certain geometric discipline to set-up coordination and measurement of components and spaces in building design (Bahri, 2009).

Modular coordination provides a coordinated and an effective system for identifying suitable joints locations. Every joint should relate to a joint reference plane. When IBS components designed, they should include product tolerances, installation tolerances and interfacing tolerances which are re- quired in production and assembly processes (Ghazali, 2006). According to Mian (2006), the concept of modular coordination was introduced in Malaysia since 1986, but the lack of experience in modular coordination and its need for precision dimensioning and accurate planning limited the use of this concept in building industries.

It was endorsed internationally that the basic module of modular coordination can be represented by the letter M which equal 100 mm. When the designers need to double dimensions on projects' plans, they multiply M by 3 series (3M, 6M, 9M, 12M etc.) to use it to design the main dimensions of building framework such as floor span, column distances, etc. While the sup modules which are 0.5M (50 mm) and 0.25M (25 mm) used in smaller dimensions (Ghazali, 2006).

According to Ghazali (2006), modular coordination is mainly based on:

- Using modules (basic module and multi-modules).
- Reference system to define coordinating spaces and zones for building elements.
- Rules for locating building elements within the reference system.
- Rules for sizing building components to determine their work sizes.
- Rules for defining preferred sizes for building components and coordinating dimensions for buildings.

The application of modular coordination principles can provide many advantages. As reported by Ghazali (2006) the main advantage of using modular coordination in building design is simplifying and clarifying building process by offering better harmony and arrangement among building industry players due to providing general language among them. Other advantages of modular coordination are limiting the diversity in components dimensions, minimizing time needed for design process and decreasing cost in manufacturing and erection processes because of using standardized components. Furthermore, it can minimize waste materials during production process. Finally, it can facilitate the industrialization of building process through increasing the use of modern technologies such as Computer Aided Design and Drafting and Computer Aided Manufacturing.

## 3.2 Closed and open building system

### 3.2.1 Closed System

A closed system can be classified into two categories, namely, production based on client's design and production based on pre-caster's design. The first category is designed to meet a spatial requirement of the client's, that is the spaces required for various functions in the building as well as the specific architectural design. In this instance, the client's needs are paramount, and the pre-caster is always forced to produce a specific component for a building. On the other hand, the production based on pre-caster's design includes de- signing and producing a uniform type of building or a group of building variants, which can be produced with common assortments of component. Such building includes school, parking garage, gas station, low cost housing, etc (Thanoon et al., 2003). According to Warszawski (1999), these types of building arrangement can be justified economically only when the following circumstances are observed:

a) The size of project is large enough to allow for distribution of design and production costs over the extra cost per component incur due to the specific design.

b) Standardization and large repetitive element in the architectural design. Automating the design and production process for the novel prefabrication system can provide the requirement of many standardized elements.

c) Sufficient demands for a typical type of building such as school so that mass production can be obtained.

d) The pre-casters should provide intensive marketing strategy to inform the clients and designers about the potential benefit of the system in term of economics and noneconomic aspects.

### 3.2.2 Open building system

An open building system is defined as the ability to use the prefabricated elements that were manufactured by different manufacturers at varied locations adopting different manufacturing processes as well as assembly of these elements to build the complete structure using assorted erection equipment (Trikha & A. Ali 2004).

Using modular coordination in industrialized building system can lead to an open building system which enables IBS components to be used at different sites without resulting any monotony or repetitiveness in building structure.

The concept of open building was firstly introduced during rebuilding years of post-war in Netherlands (Shaari, 2006). Many developed countries achieved their success in the field of open building industry. The prefabricated elements were standardized and used in the form of an open building system which can offers great benefits compared to closed system (Trikha and Ali, 2004).

Open building is an innovative approach for design and construction that enhances the efficiency of the building process, while increasing the variety, flexibility and quality of the product. In the Open building perspective, the building is viewed as a well-organized combination of systems and subsystems, each of which can be carefully coordinated to ensure a better process and product for the homeowner and a parallel positive outcome for the building professionals. Humans need to rehabilitate, remodel and change their dwellings constantly, as long as they live in. (Bensonwood, 2003).

According to Habraken (2003), open building system included two flanks which are social flank and technical flank. The social flank provided client's preferences by offering flexibility needed for adaptation of individual units over time. On the other hand, technical flank provided construction methods to install, change or remove sub-systems with a minimum of interface troubles.

To achieve the success of industrialization in Malaysian construction sec-tor, the application of open building system has to be improved. It was suggested that the development of open building system in Malaysia could be implemented through government bodies like CIDB and the Standards and Industrial Research Institute of Malaysia (SIRIM) by the development of Malaysian Standard (MS 1064) joints for IBS components (Hamid et al., 2007).

The correlation among IBS components and fitness on site are very important issues. The most common system was developed joint used in closed system protected by patent. It was supposed that CIDB through their research arm, Construction Research Institute of Malaysia (CREAM), with collaboration with Jabatan Kerja Raya Malaysia (JKR), would develop the Malaysian standard joint which supposed to be more strong, stiff, stable and ductile by the end of 2007 (Hamid et al., 2007).

### 3.3 Just in time principles

The just in time principle can be defined as a philosophy of manufacturing based on planned elimination of all waste and on continuous improvement of productivity. It was also defined as an approach with the objective of producing the right part in the right place at the right time. Another definition of just in time principle is collaborating the works between factories and the need of construction site during the construction process to provide the adequate prefabricated components at exact time.

The success of the application of the just in time principle can be obviously noticed in the industry sector. As mentioned in Wise Geek (2009), Henry Ford was the first who used Just-in-time manufacturing philosophy in Ford Motor Company. Ford's method was to buy the immediate needs of materials for manufacturing process. Later, he organized transportation of materials so the flow of the product would be smooth. This process gave him rapid turnover, decreased the amount of money tied up in raw materials and prevented the storing of materials.

Later, several cars' manufacturers adopted Ford's just-in-time manufacturing process such as Toyota which used the process with very satisfactory results like huge amounts of cash appeared as in-production inventory was built and then sold. Moreover, the time needed of building a vehicle could be reduced to about a day. Therefore, the customers were very satisfied as vehicles could be provided within one or two days and reduce the risk that many vehicles would be built and not sold. Then after, just-in-time manufacturing philosophy was adopted by many industries and businesses and achieved very successful results.

In the construction sector, the storage area at a site must be large enough to store materials and components. Furthermore, there is a need for perfect protecting the quality of prefabricated components. Providing this area at construction site is money consuming (Mian, 2006). Furthermore, the construction site could be messy and muddy. This may increase the hazards and risks of components' damage (Pheng & Chuan, 2001). Therefore, the application of just-in-time principles can save cost associated with storage of components at site, and the elements can be in the right quantities to the right place at the right time (Mian, 2006).

The successful application of just in time principle can deliver a lot of benefits to construction sector. Mian (2006) mentioned the main advantage of just in time principle which are decreasing the unnecessary activities that maximize cost without valuable results, such as the unjustified movement of materials, the accumulation of excess inventory, or not avoiding the rectification works in production process. Another profit of just in time principle is minimizing the total construction time by managing the construction processes in a right way. Just-in-time concept can be shown in Figure 3.1.

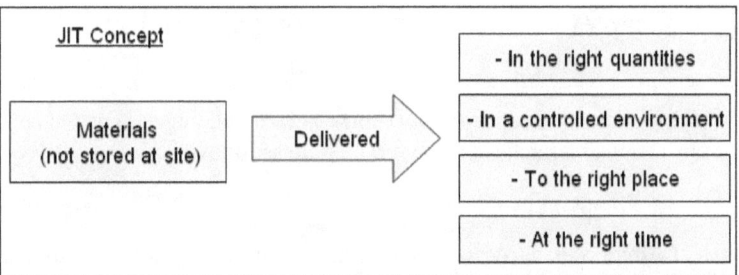

Figure 3.1: The just-in-time principle in construction industry (Mian, 2006)

Pheng and Chuan (2001) found, in their study, that the main reason of shortage in the application of just in time principles in Singapore was the lack of confidence between the contractors and IBS suppliers. The contractors were inattentive to the benefits of just-in-time principles in decreasing total cost and time. Therefore, they were not keen on implementing this principle. On the contrary, most of suppliers are ready and interested in just-in-time deliveries. The researchers suggested that the rectifications of this problem must be done by applying correct and accurate contractual agreements in the contract to promote the reliability between these two sides. It is hoped that just-in-time principle can be fully implemented in Malaysian construction industry, so that the productivity and efficiency of work at site can be improved to a greater height (Mian, 2006).

# Chapter 4: The Constraints of Industrialized Building System

In general, the constraints of industrialized building system that collected from various researches can be summed up as follows:

## 4.1 Financial problems

Despite the successful achievements in administrative structure to transform from the conventional-intensive construction to industrialized building construction, the markets are not ready, and the construction industry is still far from achieving industrialization in construction. The 2005 architects' survey revealed that 50% of respondents indicated that using industrialized building system in construction industry could not reduce total construction costs (CIDB, 2005c). The fluctuation in construction markets made the use of prefabricated building system perilous. Therefore, many developers in the advanced countries preferred to use the conventional system (Warszawski, 1999).

Cheapness of unskilled workers made the contractors prefer using the conventional method against prefabricated system (Thanoon et al., 2003). Moreover, the capital-intensive cost is still the main problem that hinders the use of industrialized building system in Malaysia (Kaur, 2009). According to Badir et al. (2002), the main disadvantage of industrialized building system in Malaysia is the need to enormous capital cost. Capital cost of industrialized building system normally includes set up plants, employing skilled and semi-skilled workers, providing machinery and moulds, and the expenditures of transportation process (Bahri, 2009).

A study carried out by Hassim et al. (2009) found that the financial failure was the most significant risk category that IBS contractors may suffer from in Malaysia. Moreover, the cost of transportation process amounts to 3%-5% of their total cost for distances not exceeding 50km-100km (Warszawski, 1999). The cost can be effective only in large projects when using repetitive-ness in design (Hong, 2006).

Furthermore, many architects and engineers are still unaware of the basic elements of industrialized building system such as modular coordination due to lack of incentives and promotions from the government in the awareness

of this innovative system (Thanoon et al., 2003). Kamar et al. (2009a) illustrated in their study that industrialized building system in Malaysia may need serious marketing and re-branding efforts.

## 4.2 Shortage in industrialized building system and modular coordination knowledge

Despite all the efforts of Malaysian government to encourage the implementation of industrialized building system, there is still lack of knowledge in this filed among construction parties. The 2005 architects' survey showed that 66% of architects acknowledged having poor knowledge in industrialized building system. Furthermore, 65% of them acknowledged having poor knowledge in modular coordination. On the suggestions' section of this survey, non-ignorable percentage (34% of the respondents) requested for more awareness and education programs on industrialized building system and modular coordination (CIDB, 2005b).

Oleiwi (2011) indicated that many academicians in the universities acknowledged that academic curricula in the related universities and institutes do not provide adequate educational courses about industrialized building system and modular coordination principles. In many universities, precast concrete design was not provided to undergraduate students (Rahman and Omar, 2006). Thus, designers and builders had more tendencies to use the familiar conventional method. The shortage of adequate awareness concerning industrialized building system among building professionals is the greatest hindrance to its successful application in practice (Warszawski, 1999). Due to shortage of knowledge in this field, many projects have been carried out with difficulties and problems. Some of these problems are the inaccurate assembly of the connections of beam-to-column and column-to-base. These problems arise due to the fact that the parties involved in the construction underestimate the important of accuracy in setting out the alignment and leveling of the bases. Basically, accurate leveling and alignment of the bases are the two most important aspects for the successful rapid erection of precast concrete components. Moreover, lack of knowledge in designing the details of ties and connections of the precast concrete components exacerbates the problem of poor connection system (Rahman and Omar, 2006). This problem can be clearly shown in Figure 4.1.

Figure 4.1: Poor connection system leads to safety problems

On the other hand, steelwork structures were also mistakenly designed to simulate the conventional reinforced concrete structural system. This concept results in exposed steel beams and columns and causes many serviceability problems such as leakage (see Figure 4.2). Rain water can easily infiltrate into the internal building through the joint between the wall and steel beam. Dampness leads to corrosion to the lighting system and the steel beam.

Figure 4.2: Steel beam should not be exposed to as in the case of concrete beam.

Therefore, knowledge and awareness in construction technology are totally important. The duty of Malaysian government becomes very important to achieve the entire industrialization in construction sector. The educational programs need to improve within one or two years (Jumaat, 2009). The government implemented great efforts to train the local construction workers. In the 2009 budget speech, it was announced to provide at least 100,000 industrial training opportunities for workers as some of CIDB duties in technical fields such as welding, management and safety. Construction workers were encouraged to obtain skills certification through accreditation and skills training conducted by CIDB and the fees was paid by CIDB for local workers (Badawi, 2008).

Through the Akademi Binaan Malaysia (ABM), the government is spending millions of Ringgits each year for the training purpose. However, these training courses are not enough if the graduates are not interested to find jobs in the construction industry due to the extremely low wages (Shaari, 2006). These training courses are also useless if the academic curriculum in universities and institutes do not provide sufficient study on industrialized building system and modular coordination. As the students studied the conventional building system for many years during their study, they would not be familiar to design in industrialized building system and modular coordination professionally after graduate, and the small duration time of CIDB training courses would not be enough (Ibrahim, 2009).

Kamar et al. (2009a) recommended in their study that industrialized building system in Malaysia might need re-structuring in Research and Development (R&D) program, re-organizing training and awareness program and providing proper incentive for information technology adoption. Further- more, it was demonstrated that the top three sources of risks of industrialized building system in Malaysia were inexperience of the contractors in IBS pro- ject, complexity in design and low contractors' performance. Theses top three sources of risks can be attributed to lack of experience of contractors and de- signers in IBS (Hassim et al., 2008). Most of contractors do not prefer the use of industrialized building system because of the high expenditure, lack of previous experience, the increment in project risks and lack of professional training regarding this new system (Kamar et al., 2009b)

Furthermore, there was a shortage in experience among workers. Many of foreign skilled workers had left the country after the widespread crackdown on illegal foreign workers in 2002. The new batches of foreign workers did not possess the required skill and they must be retrained (Thanoon et al., 2003).

Moreover, the unsuccessful initial stage of early IBS pilot projects (Pekeliling Flats in Kuala Lumpur and Rifle Range Road Flats in Penang) impressed a bad reputation among contractors about this new system. Poor quality and lack in design knowledge were the main reasons of the failure of these two pilot projects. These projects were designed by foreign designers, thus led to unknown serviceability aspects like the need for wet toilets and bathrooms (Rahman and Omar, 2006). Figure 4.3 shows the first IBS projects in Malaysia.

Pekeliling Flats, Kuala Lumpur

Rifle Range Road Flats, Penang

Figure 4.3: The first IBS projects in Malaysia

### 4.3 The issue of foreign workers

The high rising in standards of living and economic growth of Malaysian society created high demands on construction activities. As a result, a huge number of foreign workers were attracted into this country as unskilled labor doing manual jobs (CIDB, 2005c). It was reported in Migration News(1995) that in 1995 there were 553,658 foreign workers in Malaysia. According to Bernama (2006), the total number of foreign workers in 2006 rose to 1.8 million. Migration News (2011) reported that the foreign workers available in Malaysia in 2011 were about 3,300,000. This revealed that the number of foreign workers is increasing in Malaysia extensively.

It was noticed that most of the foreign workers in Malaysia were un-skilled workers. Those foreign workers were not involved in the construction industry in their countries. When they came to Malaysia, they involved in construction sector. In that case, the quality and productivity of construction production could be affected because they didn't get enough experience in construction work before (Bahri, 2009). Furthermore, the availability of cheap foreign workers in Malaysia became the reluctant force against the use of industrialized building system and made the contractors and developers prefer using the conventional method (Thanoon et al., 2003).

Most of the workers in Klang Valley were from Indonesia, Thailand, the Philippines and Myanmar. The agricultural culture of these countries influenced on workers' behaviors. These workers might never experienced with working in the construction of huge projects like condominiums and apartments (Ghani et al., 2007). Moreover, the local workers were not keen on joining in this industry because of the low wages and low emphasis on working conditions (CIDB, 2007).

Malaysia's goal of being an industrialized country by 2020 was threatened due to low-skill and low-tech trajectory (Migration News, 1995). As a consequence, the danger of the foreign workers became very clear competing with Malaysian workers. Other problems related foreign workers were money outflow to their countries, low quality works, delay in works, social problems such as fighting and quarrels which might occur between foreigners and citizens and dangerous diseases such as hepatitis and AIDS. Furthermore, the flood of foreign workers put the country in a stalemate due to growing in the essential needs especially in the important sectors such as housing, public health, transportation, education and security (Badaruzzaman, 2008). The industrialized building system was proposed in several studies as the solution to reduce the dependency on foreign workers in construction industry. By adopting the industrialized building system and reducing the use of conventional method, the dependency on foreign workers can be reduced, thus reducing billions of Ringgits currently being transferred to their home countries and reducing social problems involving those foreign workers (CIDB, 2003b). A study carried out by (Warszawski, 1999) revealed that the use of industrialized building system was 40 - 50% less labor employed than conventional method.

The issue of foreign workers was the main topic in discussion held by Building Industry Presidential Council (BIPC). It was discussed the financial problems that were caused to some contactors and developers by the foreign labor supply disruption in mid-2002. The possibility of repeating that scenario must be the driving force to make the industrialized building system the dominant method in the construction industry (Ibrahim, 2003).

CIDB took a pivotal role to reduce the dependency on foreign workers in construction industry. One objective of IBS Roadmap was to reduce the construction foreign workers percentage from 75% of total construction workers in 2003 to 55% in 2005 and 15% in 2009, especially those involved in wet- trades such as carpenters and plasterers (CIDB, 2003b). The use of industrialized building system for many years will bring the benefit to Malaysian society both economically and socially (Badaruzzaman, 2008).

### 4.4 Technical limitations

The fragmentation and diversity in the construction industry sector led to the difficulty of organizing IBS planning stage, which needs consensus among parties (Thanoon et al., 2003). Furthermore, industrialized building system is limited to changes which might be required in the building over its economic life especially in small span room size (Warszawski, 1999). More-over, there was a weakness of connection and jointing methods used in this system. This issue was very critical and sensitive to the error and sloppy work (Warszawski, 1999). However, precast technology has since improved and the problems of joints have been solved (CIDB, 2005a).

### 4.5 Transportation limitations

The transportation of prefabricated components from the manufacturing yards to the construction sites is an important stage in IBS projects. Transportation process had its limitations such as damages during transit IBS components like overturning, sliding, slipping or twisting of the components. The other limitations include hazards during removing the units from the de- livery vehicles. Special care must be taken to prevent these damages when transit the elements. Furthermore, the access consideration for trucks and cranes include traffic control, street access and unrestricted site operation are important issues that must be taken into consideration when transport IBS components. The roadway and approaches should be level, firm, well compacted and stable. Sidewalks, underground utilities and tanks and overhead lines should be protected also (Trihka & Ali 2004).

Difficulties to access to site and difficulties to transport big components from the factories to construction sites may hinder the ability to use industrialized building system for a particular site. Moreover, Badir et al. (2002) mentioned that transporting IBS components could be delayed due to bad weather, and that was the third main reasons for delay in IBS projects completion.

# Chapter 5: Studies on Industrialized Building System in Malaysia

## 5.1 Studies on industrialized building system in Malaysia

Several studies have been conducted to study the industrialized building system in Malaysia. In 2003, a study carried out by CIDB (2003a) to identify the extent of IBS usage among contractors. Questionnaires were sent to all contractors listed under the G7, G6 and G5 CIDB's categories (5172 companies). However, the response rate was small (only 4% of the total companies responded the questionnaire). The results indicated that the usage of industrialized building system in Malaysia becoming more popular and about one third of the projects completed from 1998 to 2000 used industrialized building system in one form or the other. However, the usage of this system in certain areas was still quite low and some parties, especially the clients and government agencies need to be informed on the benefits, and the correct applications of the systems.

In 2005, CIDB conducted another study to survey the architects' opinion, based on their experience, on the performance of IBS projects. Questionnaires were sent to all architectural firms (1384) registered by the Board of Architects Malaysia. However, only 7% of the companies responded the questionnaires. The findings demonstrated that by using the industrialized building system, faster, safer, cleaner and neater construction can be gained, though not the cost. The findings also reflected the poor experience of architects in industrialized building system and modular coordination. It has been recommended in this study that it would be better to execute this type of study through personal interviews even though it would require more time and re- sources (CIDB, 2005b).

Kadir et al. carried out another questionnaire survey in 2006 (Kadir et al., 2006), to compare between the conventional method and the industrialized building system in terms of actual labour productivity, structural cost, crew size, and cycle time. A total of 100 respondents participated in this study. Analysis of variance (ANOVA) test indicated that there was significant difference between the conventional system and the industrialized building system in actual labour productivity, crew size and cycle time, and insignificant difference in term of structural construction cost. However, the sample of this study was restricted to the residential projects only.

Another study carried out by Wee (2006) indicated that the acceptance of Acotec wall panel - which is considered as a type of industrialized building panels- among the respondents was just at average level. Acotec is Advanced Construction Technology wherein the lightweight precast hollow core panels made of normal Portland cement, sand and an expanded lightweight aggregate are used for non-load bearing partitioning. The sample of this study was the local contractors and manufacturers. 57% of the targeted sample responded the questionnaire. The findings revealed the acceptance of Acotec wall panel application was just at the average level. However, this study was limited to one type of industrialized building system named Acotec wall panels.

In 2008, a survey carried out by Basri (2008) to identify and rank the success factors and barriers in the implementation of industrialized building system in Malaysia. 50 questionnaires were mailed and 27 were received. The respondents were site workers, site engineers, site supervisors, foremen, management staff and production manager of the precast factory. The study was limited to residential projects only. The study found that initial cost was the first barrier in the implementation of industrialized building system in Malaysia, followed by lack of knowledge among construction parties.

Another study carried out in 2008 by Idrus et al (2008) to clarify the reasons for the cold reception of industrialized building system in Malaysia. The survey research methodology has been used for this study. A target population of 180 respondents was set, consisting of 90 contractors, 30 developers, 30 architects and 30 engineers. The study population was further confined to only the northern states of Peninsular Malaysia, namely, Perak, Penang and Kedah. Only 19.4% of the respondents answered the survey. The results of this study indicated that industrialized building system need high cost due to lack of economy of scale in IBS projects and business monopoly by the small number of IBS producers in Malaysia. Furthermore, a large faction of the industry felt that the technology transfer in industrialized building system was unsuccessful. They claimed that there are insufficient IBS guidelines and unclear standards for industrialized building system in Malaysia. Moreover, it is hu- man nature to resist change from the more familiar which is conventional method to the less familiar, industrialized building system. Also, the conventional in-situ system is still attractive as it is cheap. The results also showed that the present incentives given by the government to promote industrialized building system are inadequate to a large faction of the industry.

In 2010, Kamar et al. (Kamar et al., 2010) carried out another survey to identify the Critical Success Factors (CSFs) of industrialized building system in Malaysia. All the G7 contractors who registered under CIDB (355) were included in this study. The responses were only 37 companies (10.42%). It was inferred that the quality and speed of construction were the two major driving forces for the use of industrialized building system in Malaysia. The cost and high capital investment needed to implement industrialized building system were still the major barriers hindering this system.

From all the above studies, it can be concluded that industrialized building system in Malaysia is still in the early stages. High initial cost is the main constraint that hinders the usage this system widely in the Malaysian construction sector. On the other hand, the main advantages of this system such as time factor, better quality besides getting cleaner and neat site are to be considered imperative in the Malaysian construction area.

## 5.2 Recent studies on industrialized building system in Malaysia

The most recent study on the industrialized building system in Malaysia carried out by Oleiwi (2011) to evaluate the advantages and constraints besides receiving suggestions to improve this system in Malaysia. The first method that has been considered in this study was distributing questionnaire which were limited to the companies that registered in IBS Center database. These companies were categorized under three types which are IBS manufacturers, IBS consultants and IBS contractors. The second method was conducting face-to-face interviews with IBS manufacturers, IBS consultants, IBS contractors (that registered in IBS Center database), the interested academicians in the universities and the local authorities from CIDB Malaysia.

To overcome the limitations of the previous studies, the present research included all the above important parties that interested in this system. The information was collected by qualitative and quantitative methods mentioned earlier to increase the strengths of each method and reduce their weaknesses. Furthermore, all the types of IBS components were included.

### 5.2.1 The questionnaire

In this research (Oleiwi, 2011), a questionnaire form was designed as simple as possible for easy understanding by the respondent and to minimal filling time. The questionnaire form produced in two phases. First, the pilot survey was developed based on the reviews of the related literature. Then it was explained to experts from different universities to pre-test the survey and prevent any possible confusion or difficult questions. The feedback and the comments of the experts to improve the survey were reviewed and treated respectively. After the first phase was over, the final revision of the questionnaire was delivered and distributed to IBS manufacturers, consultant and contractors registered in IBS Center, Kuala Lumpur, Malaysia using IBS online survey system.

In traditional surveys "paper and pencil" method was used to collect primary data in different academic research fields for several years (Roztocki & Lahri, 2002). But this traditional approach always had the disadvantage of limited response rate and spending long time in distributing and returning the survey paper (Roztocki, 2001). It was proved that Web-Based surveys could be considered as a serious alternative to the traditional "paper and pencil" for- mat in all academic research fields. The popularity of online data collection methodology has increased recently in the advanced countries (Roztocki & Lahri, 2002). The online survey system can provide opportunities to make the surveys in more efficient and effective way than the traditional approach. Also, only low response rate is expected for mailed surveys also. Mail surveys do need a long time, cost and labour. By using online survey system, all these limitations could be circumvented (Roztocki, 2001). It is expected that the importance of internet-based surveys will increase for academic data collection due to the ability of this approach to deal with problems and the non- ignorable advantages (Roztocki, 2001).

The questionnaire of Oleiwi (2011) research consisted of five parts, which were as follows:

- The first part contained questions on general information about companies.

- The second part contained questions about the details of IBS projects done by companies.

- The third part contained questions on the advantages of using IBS in Malaysian construction sector.

- The fourth part had questions about the constraints of IBS in Malaysian construction sector.

- The fifth part contained questions about the suggestions to improve IBS in Malaysia.

An introduction was presented with the questionnaire in the beginning to introduce the researcher to the objectives of the study. The web-page that included the first part of the questionnaire is shown in Figure5.1.

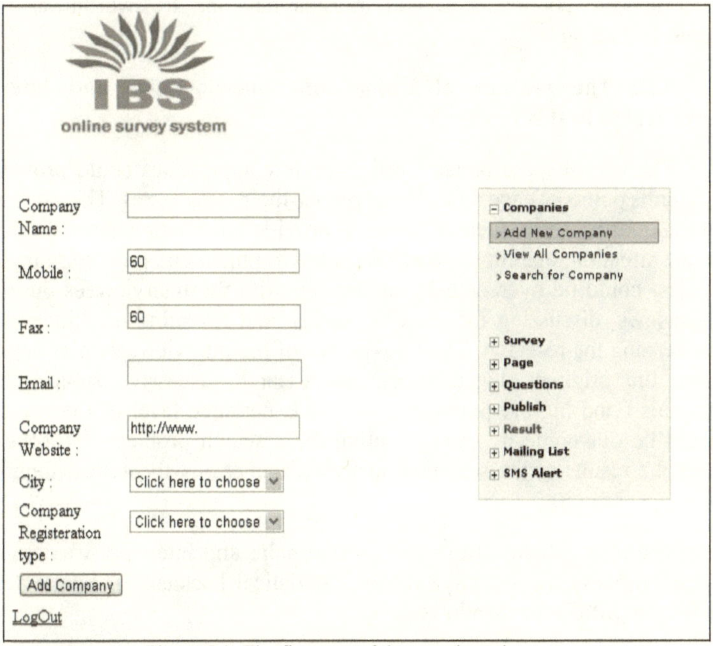

Figure 5.1: The first part of the questionnaire

### 5.2.2 The interviews

The interviews enabled the researchers to probe deeper into the topics presented in the questionnaires. The interviewees provided deep information and specific examples of their experience in IBS field that supported the research issue. Seven steps were followed in conducting the interviews. The first step was formulating study questions, followed by preparing a short interview guide and selecting key informants. Then the interviews were conducted, and adequate notes were taken from the interviewees. The data obtained was analyzed and finally reliability of the data was arrived at.

Although the questions of the interviews were the same as in the questionnaire, it contained interviewees' comments to probe further information regarding the topic and in particular the underlying reasons for the responses. According to Matveev (2002), researchers should include quotations from different participants to add transparency and trustworthiness to their findings. Therefore, this study included some of interviewees' comments to add transparency and trustworthiness to the research findings.

### 5.2.3 The reasons of using both questionnaire and interview approaches in this research

The use of questionnaire and interview approaches could provide the strengths points of each method and reduce their weaknesses. The weaknesses of the questionnaire, such as failure to provide information about the context of the situation, inability to control the environment and pre-determined out-comes, could be overcome by interaction with the interviewees during the interviews, discussing the research issues, and revealing the hidden issues concerning the research. The weaknesses of the interview, such as departing from the original objectives of the research, excessive subjectivity of judgment and high requirements for the experience level of the researcher could be overcome by clearly stating the research problem, crosschecking with the results of the statistical analyses, and strong theoretical foundation of the re- search.

The data gathered from the questionnaire and interview was arranged, coded, processed and analyzed using Statistical Package for Social Science software (SPSS) for Windows.

The study revealed that time reduction, environmental advantages and high quality were the most valuable advantages of industrialized building system in Malaysia. The finding also indicated that lack of experience, high cost and payment delay were the most critical constraints for this system in Malaysia. Moreover, improving the academic curricula concerning industrialized building system and modular coordination was the first recommendation that was put forward to improve this new building technology in Malaysia.

## Chapter 6: Well-known IBS Projects in Malaysia

The Malaysian government has been highly committed the implementation of industrialized building system. After the 1990s, Malaysian construction witnessed such a significant improvement in terms structural performance and architectural aspects. One of most important reasons is the advanced knowledge in Computer Aided Design. With this knowledge, IBS projects can be designed and visualized analytically before construction process and hence the rectifications to the component design can be done before the manufacturing process. The 3-dimensional drawings can provide accurate component dimensions to ensure build-ability. These 3-dimension drawings can also provide simulation of erection and construction procedures. Feasibility studies on the different building systems can be performed without incurring much cost. Problems during construction can also be observed and predicted (Rahman and Omar 2006).

As a result, to this development, many IBS successful projects has been constructed throughout Malaysia. Some of the well distinguished IBS projects were explained briefly as follows:

### 6.1 Serdang Hospital – IBS hybrid of precast concrete and steel structures

Besides high-rise buildings, the usage of steel elements is also popular with the construction of universities, colleges, schools, hospitals and commercial complexes. Undoubtedly, structural steel offers greater freedom and flexibility to the designers, rapid construction for the contractors and faster returns on investment (ROI) to the owners. Moreover, by using this type of building, it is possible to achieve time saving of up to 30 percent, which leads to save labour, material and cost. This was the case with the new Serdang Hospital.

Serdang Hospital is a government-funded multi-specialty hospital located in the state of Selangor, near Putrajaya, the Malaysian federal government administrative center. The location of the hospital borders the South Klang Valley Expressway (SKVE) to the east and the medical faculty of Universiti Putra Malaysia to the west.

The construction of Serdang Hospital started operation on December 15, 2005. This project was built with the purpose of serving the roughly 570,000 population of Serdang, Putrajaya, Kajang and Bangi. It has a total area of

129,000 square meters and has 20 operating theatres, 19 wards and 620 beds. The total construction cost that has been expended for executing the hospital was RM 690 million.

The hospital is one of the electronic hospital networks planned by the Malaysian Government to provide better health care services to Malaysian, in line with its vision of becoming a developed country by the year 2020. It was designated as the reference center for cardiology, cardiothoracic, urology and nephrology surgery. The hospital provides affordable, quality treatment for heart patients from the lower income group. The hospital also serves as a teaching hospital for medical students from the Faculty of Medicine, Universiti Putra Malaysia (UPM) and Cyberjaya University College of Medical Sciences (CUCMS). The hospital was also listed as one of the successful IBS projects that have been built in Malaysia in 2005 (CIDB 2005c).

Serdang Hospital is the first hospital in Malaysia to use aluminum coating to give it a shiny exterior look. Thus, the hospital can be called a "hospital of the future". The hospital is surrounded by a park measuring 45 acres (180,000 m2). The landscape of the hospital is also known to be therapeutic to the patients.

Figure 6.1: The aluminum coating of Serdang Hospital's facade

It was the first hospital in Malaysia to be built using the hybrid IBS -steel and precast concrete structures. Instead of the concrete beams and columns, more than 6,100 tons of steel elements have been used to build the main frames.

Lightweight blocks were used for partitions to reduce the dead load of building. Besides that, aluminum claddings were also extensively used at the façade to produce a modern look. All of these have contributed to smaller structural steel section, thinner floor slabs and less piling works.

The main structural components for the main 500m long building consist of structural steel beams and columns. The structural floor slabs were constructed by using permanent concrete formwork (Half Slab) panels that were later topped with in-situ concrete. For this project, all the steel elements were prefabricated and cut to size offsite. The steel columns and beams were produced in the Perwaja factory in Gurun, Kedah and brought to be cut to size in a factory in Puchong. The permanent concrete formwork was produced in a factory in Seremban and delivered to site based on just in-time principles. Due to all of the above, material storage and wastage were kept to a minimum; and made the site neater, safer and easier to manage.

Rapid speed of construction has been achieved in this project due to using permanent concrete formworks and structural steel elements. The erection of steel beams and columns and the installation of the concrete formwork panels took less than four months for the whole area of 130,000 $m^2$; or about 1,200 $m^2$ per day.

This project proves that, by carefully selecting and balancing the structural systems, hybrid precast concrete and structural steel buildings can be optimally designed to benefit from the economical use of concrete, speed and flexibility of structural steel elements. Significant savings in both cost and time can then be achieved; helping the government to offer the ultra-modern hospital services to the public earlier as well as providing faster returns on investment to the other tenants occupying the hospital blocks (CIDB 2005c).

### 6.2 Government apartments in Putrajaya - Precast concrete walls

The apartments located in the state of Selangor, Putrajaya, the Malaysian federal government administrative center. The project has been planned by the Malaysian Government to provide suitable accommodation for the employees who work in government departments in Putrajaya. The project marked a notable significance as it supports the Government's efforts in promoting the use of industrialized building system (IBS) technology in construction. The apartments, five blocks of 16 and 19 storeys, are complemented with a total of 548 apartment units and amenities such as car parks, playgrounds, multi-purpose hall and praying hall. Precast concrete walls were used in this project.

Figure 6.2: Government apartments in Putrajaya

### 6.3 Tuanku Mizan Zainal Abidin Mosque - Steel framing system

The Tuanku Mizan Zainal Abidin Mosque, or Iron Mosque is the second principal mosque in Putrajaya after Putra Mosque. The construction was begun since April 2004 and was fully completed on August 2009. It was officially opened by Tuanku Mizan Zainal Abidin on June 11, 2010.

The mosque was built to cater to approximately 24,000 residents including the government employees working around the city center. The total construction cost was RM 208 million (about US$ 55 million).

The mosque also called "Iron Mosque" due to the construction material that was used in this mosque. The mosque employs "Architectural Wire Mesh" imported from Germany and China to clad the 20 façade-high and eight-meter-wide arched windows of the Mosque. In this connection, the metal mesh did not only achieve aesthetic requirements, but also protected its congregation dependably against sun and driving rain. The simultaneous filtration of the air flow makes the stay enjoyably cool. Thus, the Mosque, which is located right next to the newly installed Putrajaya lake, offers its visitors protection and calmness for their ritual location all year long.

Figure 6.3: The facade of Tuanku Mizan Zainal Abidin Mosque

The structure of the mosque was executed using steel framing system which is one type of industrialized building systems. The mosque was built using approximately 6,000 tons of steel, or seventy percent of the entire building.

Figure 6.4: The construction process of the mosque

Figure 6.5: steel framing system that used in the mosque

The architectural style of the mosque is the Islamic modern style. The main entrance is reinforced with Glass Reinforced Concrete to increase the integrity of the structure and uses fine glass to create an illusion of a white mosque from afar.

Figure 6.6: The main entrance of the mosque

The path towards the mosque crosses a skyway known as the Kiblat Walk which stretches an area of 13,639 m².

Figure 6.7: Kiblat Walk

There is also a Mihrab wall which is made of 13-meter-high glass panel with 2 verses. The Mihrab wall is designed so that no light will be reflected, creating an illusion that the verses are floating on air.

Figure 6.8: Mihrab wall

## 6.4 Kuala Lumpur International Airport (KLIA) - Steel roof structure

Kuala Lumpur International Airport (KLIA) was built in the jungle with features that allow flexibility for future expansion. The airport is built on 10,000 hectares which makes it one of the largest airport sites in the world. The airport was completed in four and a half years with round-the-clock construction work which make it the fastest airport ever built. The project was undertaken by an international workforce of 25,000 people who is considered as a largest number of workers for a Malaysian project. The total cost of the project was about US$3.5bn. The airport commenced full operations on June

28, 1998.The airport is approximately 75 kilometers south of Kuala Lumpur's city center and surrounded by four main cities of Kuala Lumpur, Shah Alam, Seremban and Malacca.

The airport has two parts, which are connected by a monorail. Steel roof structure has been used as an industrialized system in the roof structure of the Main Terminal Building (MTB) of Kuala Lumpur International Airport. The main terminal building roof structure is made up of keel trusses connected to the top of the conical concrete columns. The main trusses are connected to the concrete columns by hinge bars.

Figure 6.9: The main terminal building roof structure

The structure was erected using 150t crawler cranes from the ground. The total job consisted of 9305 tons and erected in 8 months. Kuala Lumpur International Airport is a destination in itself. It is unique because it has within its boundaries all that are needed for business, entertainment and relaxation.

Kuala Lumpur International Airport has a futuristic look. Everything is very spacious and serene. A great amount of glass, white colors and shiny floors give the sense of walking in space station.

Figure 6.10: The shiny interior look of KLIA

The effort has been made to create a homely airport with a serene environment combined with high technologies attractions. Nature and greenery are part of the airport in line with the "airport in the forest and forest in the airport" concept. The natural environment of the airport is transformed to functions and activities that continue to enhance the nature.

Kuala Lumpur International Airport is close to Malaysia's Administration Capital - Putrajaya, the country's new administrative center, is within 20 minutes away from KLIA. The Pan Pacific Hotel Kuala Lumpur International Airport is available within walking distance from the airport terminal building for other airport users to stay and enjoy all the facilities.

### 6.5 Kuala Lumpur Convention Centre

The Kuala Lumpur Convention Centre is a purpose-built convention and exhibition facility strategically located in Kuala Lumpur, Malaysia, and part of the Kuala Lumpur City Centre precinct, known as KLCC. The convention centre designed to be a "city within a city" this 40-hectare (99-acre) site includes the 50-acre (20 ha) KLCC Park and the PETRONAS Twin Towers.

The purpose of Kuala Lumpur Convention Centre is hosting conferences, exhibitions, seminars, meetings and entertainment. The Centre is placed alongside public transportation and hotels. In February 2007, The Centre became the first convention center in Asia to achieve 'Benchmarked' status by Green Globe, the global benchmarking certification, and improvement system for sustainable travel &tourism. The underwater park (Aquaria KLCC), 60,000 square feet (5,600 m2) is located beneath the convention center.

The convention center has two auditoria; Plenary Hall and Plenary Theatre (seating 3,000 and 500 respectively); a Grand Ballroom, three Conference Halls, a Banquet Hall, five Exhibition Halls (9,710m2 of column-free exhibition space on a single level), 20 meeting rooms plus the latest in 3Gtelecommunications and digital audio-visual facilities.

Figure 6.11: Kuala Lumpur Convention Centre

The roof of Kuala Lumpur Convention Centre was executed using a combination of prefabricated steel roof truss with composite steel deck flooring system.

The distinctive bird-like design of the roof demanded specific requirements especially in terms of acoustic performance, ease of maintenance and durability. The specific challenges faced in the development of the roof construction ensured that the roof is long lasting and that it can provide the best indoor acoustic climate.

ROXUL Hardrock 80 on a flat roof membrane construction was used to diffuse the impact of rain noise on the roof as well as enabling foot traffic for easy installation and maintenance work. Hardrock also offers the added benefits of good thermal performance and fire safety properties.

Figure 6.12: The roof of Kuala Lumpur Convention Centre

## 6.6 Other IBS projects can be shown as below (IBS Center 2012)

1. Custom, Immigration & Quarantine Complex, Johor Bahru

Figure 6.13: Custom, Immigration & Quarantine Complex, Johor Bahru

- Function of Building: Custom, Immigration and Quarantine Complex and Government Offices.
- Location: Johor Bahru
- IBS System: Precast concrete beams, columns and hollow core slabs
- Owner: Public Works Department, Malaysia.

2. Open University formerly known as JPA, Kuala Lumpur

Figure 6.14: Open University formerly known as JPA, Kuala Lumpur

- Function of Building: Campus Building
- Location: Kuala Lumpur
- IBS System: Precast concrete beams, columns and hollow core slabs
- Owner: Public Service Department, Malaysia (JPA).

3. Telekom Tower, Kuala Lumpur

Figure 6.15: Telekom Tower, Kuala Lumpur

- Function of Building: Office
- Location: Jalan Pantai Baru, Kuala Lumpur.
- IBS System: Steel structure for the sky garden and top part of the building
- Owner: Telekom Malaysia Bhd.

4. Water Sports Complex, Putrajaya

Figure 6.16: Water Sports Complex, Putrajaya

- Function of Building: Water Sport Complex
- Location: Precinct 6 Promenade, Putrajaya.
- IBS System: Tubular steel, steel decking for the floor system
- Owner: Perbadanan Putrajaya.

5. KL Sentral Station, Kuala Lumpur

Figure 6.17: KL Sentral Station, Kuala Lumpur

- Function of Building: Railway station, Commuter station, LRT station, KL Sentral – KLIA Airport Rail Station, Shopping Complex, Monorail Station
- Location: Brickfields, Kuala Lumpur
- IBS System: Steel roof structure, precast hollow core slabs.
- Owner: Kuala Lumpur Sentral Sdn. Bhd.

6. Government School, Johor Bahru

Figure 6.18: Government School, Johor Bahru

- Function of Building: School Building
- Location: Taman Mutiara Rini, Johor Bahru
- IBS System: Steel beams and columns.
- Owner: Ministry of Education, Malaysia.

# Chapter 7: Suggestions to Improve Industrialized Building System in Malaysia

The very important question one might ask is: are there any ways to improve the industrialized building system in Malaysia as an example of the countries which have adopted this type of building system? The author has collected some suggestions from the architects, engineers, contractors, academicians in the universities and local authorities who involved in industrialized building system during the lengthy face-to-face interviews. These suggestions can be summarized as follows:

1. Developing the educational programs. During the interviews, the interviewees mentioned that there were short courses for industrialized building system and modular coordination in the universities; however, these courses were not enough (Tahir 2009). The graduated engineers and architects could not start working in IBS field easily because they spent several study years designing in conventional method (Mahmud, 2009). The educational programs need to improve within one or two years (Jumaat, 2009).
2. Increasing research and development process in universities & institutes to get a deep knowledge in the field of industrialized building system (Ibrahim, 2009).
3. Volume and cost issue. Badaruzzaman (2008) mentioned that achieving cost saving in industrialized building projects cannot applied successfully if there is no repetitiveness in the design of these projects such as schools, hospitals and housing projects. The architects are responsible for this problem and they must examine the available design to achieve cost and time saving (Ibrahim, 2009).
4. There is a need to look at the long term of cost issue (life cycle of cost analysis); however, most contractor looked at the initial cost of IBS projects only. The role of the government is to provide the initial cost and increase the incentives to encourage the developers to convert to this system. The use of industrialized building system for many years will bring high benefits for Malaysian society both economically and socially (Badaruzzaman, 2008).
5. The architects are the leaders of the construction process; therefore, they must open their minds towards industrialized building system. The government has to increase the awareness of industrialized building system among architects (Ibrahim, 2009).

6. Providing initial capital cost to the developers from the government. The contractors and the consultants who want to establish their own factories and produce IBS components need grants and monetary supports from the government. This developing in the activities of the contractors and the consultants can increase cost benefits for them and make the use of industrialized building system more reliable and cost effective (Kishlan, 2009).

7. Adopting intensive training courses for IBS stakeholders. Construction Industry Development Board (CIDB) provides training courses for IBS manufacturers, consultants and contractors through Akademi Binaan Malaysia (ABM). However, these courses are not enough due to their short time (3 or 4 days). All IBS stakeholders must be exposed to intensive training courses to be able to work in the field of industrialized building system (Ibrahim, 2009).

8. Jaafar (2009) mentioned that the use of industrialized building system in private projects has to be compulsory to enforce the developers to shift to this system. However, other interviewees have different points of view. Ali (2009) mentioned that Malaysia is free country and it is better to encourage the developers in private sector to use industrialized building system by increasing the incentives and grants.

9. One of the most important problems is the abundance of cheap foreign workers. The government should legislate regulations to limit the number of foreign workers who enter the country every year. This will decrease their numbers and increase their wages and as a result, enforce the developers indirectly to tend to industrialized building system (Jaafar, 2009).

10. Legislate regulations to enforce IBS suppliers to deliver IBS components on need by applying just-in-time principle (Bahri, 2009).

11. Developing payment method in IBS projects. Banico (2009) mentioned that the contractors suffered from delay in payment in IBS projects. Payment method that they normally use to follow in conventional building system is not suitable for IBS projects. The empowered persons in CIDB focused on developing a new payment form for IBS projects to bind the clients to pay at exact time (Bahri, 2009).

12. Improving the application of open building system. The system that is used in Malaysia is closed system which has its limitations of not matching with the elements of other companies. Open building system provide using the prefabricated elements from different manufacturers at varied locations without any mismatching problems (Bahri, 2009).

13. Applying quality supervision from the government in all IBS factories to insure that the quality of IBS components matches the standardization

(Bahri, 2009).

14. Increasing the production of IBS components to enrich the markets, so the prices of these components will be decreased (Ibrahim, 2009).
15. Using software systems in different stages of IBS projects (Tahir 2009).
16. Achieving good communication and periodic meeting among projects' parties. During the interviews, many contractors complained that they need to contact with the designers directly to discuss projects' designs. The solution of this problem is to achieve periodic meeting among between the designers and the contractors (Abdullah2009).

# Appendix

In order to grasp a clear picture of the situation of industrialized building systems in Malaysia from the experts working in this field, the author has conducted the following interviews which are summarized as follows:

**The interview with Dr. Mazlan Mohd Tahir, Architectural department, Universiti Kebangsaan Malaysia (UKM )**

**First part: Respondent's information**

- Name: Mazlan Mohd Tahir

-Specialization: Architect

**Second part: IBS Information**

1- How many projects have you handled in general?

- I've handled many projects.

2- How many projects have you handled using industrialized building system?

- I've designed more than 15 projects using industrialized building system.

3- How long has your experience in the field of industrialized building system?

- I've started working in IBS projects since 1998 in Canada.

4- What types of IBS components do you used in your designs (Pre-cast Concrete Framing, Steel Formwork Systems, Steel Framing Systems, Timber Framing Systems or Block Work Systems)?

- In most designs, I used pre-cast concrete system.

5-What types of IBS projects have you designed (Residential, institutional, Social, Industrial)?

- Most of IBS projects that I handled were institutional projects like schools, universities and military projects in Canada & USA.

**Third part: The advantages of industrialized building system in Malaysia**

**1-Cost**

- Does the total cost of the project go down when industrialized building system is used?

- Yes. By using industrialized building system, the total cost of the project can be reduced, but the problem is that the initial cost is very high.

**2-Time**

- Does the total construction time reduced when industrialized building system is used?

- Yes. The total construction time can be reduced to approximately less than 20% compared to the time needed if conventional method is used for the same project.

**3- Number of workers**

- In your view, does industrialized building system capable of reducing the number of workers in factories and at construction site?

- Yes. By using industrialized building system, the dependency on workers will be reduced due to using machines in the factories and at construction sites.

- What are the problems of using foreign workers in construction industry?

- There are many problems that come with foreign workers like dangerous diseases and social problems

- What happened in 2002 concerning foreign workers?

- In 2002 there was lack in skilled and semi-skilled workers when the government expelled many of them due to their illegal interring inside Malaysia

## 4- Quality

- Does the use of industrialized building system capable of enhancing quality of buildings?

- Yes. The quality of buildings is increased when using industrialized building system due to quality control inside the factories and employing skilled and semi-skilled workers.

## 5-Safety

- Can industrialized building system provide safe environment to the workers?

Yes. The use of industrialized building system is very safe in factories and at site. The accidents are very seldom.

## Forth part: The constraints of industrialized building system in Malaysia

### 1-Lack of experience:

-Does the academic curricula in the universities provide adequate educational courses about industrialized building system and modular coordination?

- No. The undergraduate students receive insufficient courses about industrialized building systems and modular coordination. The materials of the courses depend on the initiatives of lecturers.

- Are the architects in Malaysia unable to design IBS projects?

- The architects are not familiar to design IBS projects due to shortage in training and education. They need good training courses.

- Is there shortage in the experience of industrialized building system among engineers?

- Yes, there is shortage of IBS experience among engineers in Malaysia due to shortage in education and training programs

- What about the experience of contractors?

- Many contractors don't have adequate experience to handle IBS projects because they used to work in conventional method

## 2- Cost constraints

-What are the reasons that make the use of industrialized building system is not attractive choice to the clients?

- The high initial cost is the main reason. Moreover, the conventional method is more popular and it's difficult to change from a popular method to a new method.

- What are the main reasons that cause high expense in IBS projects?

- The main reason is the high initial cost that IBS projects need. This cost includes the expense of establishing a factory, buying machineries, employing skilled and semi-skilled workers, etc.

- Are the swing of markets' demands play a role in the cost of industrialized building system?

- Of course.

## 3- Technical constraints:

-Do you think that industrialized building system has bad reputation among architects and customers?

- Somewhat.

-What are the reasons behind that?

- The first two pilot projects that used industrialized building system were not successful. The first pilot project was in Jalan Pekeliling in Kuala Lumpur, and the second pilot project was in Jalan Rifle Range in Penang. Unfortunately, the two pilot projects faced the problem of leakage and moisture. Thus, they represented bad experience of industrialized building system in Malaysia. However, this problem has been solved due to the development in IBS technology.

- Are just-in-time principles applied in Malaysia?

- No.

-Is there any problem that was caused by thermal expansion in IBS projects?

No. There is no problem because the deference in temperature is very small in Malaysia and doesn't cause any problem.

- Are the materials that normally used in producing IBS components suitable to Malaysian environment?

-Yes. The materials are suitable.

-Evaluate the performance of precast concrete systems in earth-quakes

   - Well performance          - Good          - Poor

- Good

- Why?

- Earthquakes is very simple in Malaysia.

- Are the IBS projects capable of rehabilitation?

- The structure elements cannot be rehabilitating, while the secondary elements can be rehabilitating.

**4- Transportation constraints**

- Is there any problem in transporting IBS components?

- There is no problem anymore because the limitations in components' size have been developed to control traffic accidents. However, this may limit the architectural creativity.

- Are there any traffic accidents that happened during transporting IBS components?

- Very seldom

**5-** Aesthetical constraints

-Does the industrialized building system limit the architectural creativity?

- No. There are many creative projects that built using industrialized building system all over the world, and they are very creative.

- Do you think that the increment of usage of industrialized building system in future will bring the similarity and boredom to our towns?

- No. If there will be creative designs, there will not be any boredom or similarity.

**6-** Payment constraints

- What are the main reasons of payment problems in IBS projects?

- Delay in payment is the main problem

-Is the conventional payment method suitable for industrialized building system?

- No. The payment in conventional method depends on the executed work; however, the payment in industrialized building system depends on the need of the work like providing the components from the factory, buying machineries for installing IBS components, etc.

**7- Production constraints**

- What are the main problems that may encounter IBS production?

- Some factories are not covered. The production process is exposed to weather effects and this may affect on the quality of IBS components.

- Are there other problems happened in any stage when using industrialized building system (your additions)?

1- The contractors do not want to change to industrialized building system because they used to work in conventional method.

2- Lack of experience. The contractors do not have enough knowledge in industrialized building system

3- In fact, there are no real problems, and IBS is very simple

**Fifth part: Suggestions to improve industrialized building system in Malaysia**

- Do you have any suggestions to improve the performance of industrialized building system in Malaysia?

My suggestions to improve industrialized building system in Malaysia are:

1- Increasing the awareness of industrialized building system among all the construction different parties.

2- The architects must open their minds

3- Local workers can be trained

4- The government must be the enforcer and the implementer to use industrialized building system in public and private projects

5- Developing the educational programs in architectural and civil engineering departments in universities is very necessary.

**2- The Interview with Ar. Nur Akmal Goh Abdullah, architectural department, Universiti Kebangsaan Malaysia (UKM ).**

**First part: Respondent's information**

Name: Ar. Nur Akmal Goh Abdullah

Specialization: lecturer (architect)

**Second part: IBS Information**

Experience in IBS: 16 years in IBS in Singapore

Type of IBS components used in designs? Pre-cast Concrete

Types of IBS projects: Residential, institutional, Social, or Industrial?

- Institutional (schools, universities…)

**Third part: The advantages of industrialized building system in Malaysia**

In your point of view, what the advantages of industrialized building system that Malaysian construction has achieved?

1- By using industrialized building system, the total time of projects can be decreased.

2- The quality of buildings can be enhanced.

3- This system is very safe during construction process and the accidents are very seldom.

4- Singaporean construction industry has achieved cost reduction due to mass production; however, the cost of industrialized building system in Malaysian construction industry is still expensive.

**Forth part: The constraints of industrialized building system in Malaysia**

In your point of view, what the constraints of industrialized building system that Malaysian construction still suffer from?

1- Lack of experience is the main problem. The contractors don't have adequate experience in industrialized building system. The engineers, architects and workers also don't have adequate experience in this field.

2- High initial cost to start an IBS projects is also an important constraint that have to be taken into accounts. If this problem didn't solve, the use of industrialized building system in Malaysian construction industry will be very limited.

3- Foreign workers are still in large numbers in Malaysia. The problems of the foreign workers are that they are unskilled, and they may bring social problems.

4- Just-in-time principle is not applied in Malaysia due to the delay in payment and lack of experience. This problem could increase the total time of the project and cause delay in projects and cost increment.

5- Transportation problems: Transportation regulations put some limitations to the dimensions of IBS components to avoid accidents that may occur during transporting IBS components. Therefore, the accidents are very seldom.

6- Storage constraints: during the construction of monorail train project in Kuala Lumpur, we faced a problem in storage process because the panels were very big and the factories and their storage area were very far from the project. this problem could be solved using just-in-time management.

7- Machinery: there is no problem in machinery, but the problems are in the experience of the workers to operate the machineries.

8- Boredom and similarity of industrialized building system?

Yes and no. There must be a very good understanding about the joints of industrialized building system and the modular coordination concept to explore architectural creativity and to be more innovative.

9- Earthquake: there is no problems happened because earthquakes in Malaysia are very simple.

10- Materials: in Malaysia, they use the local materials so there is no cost added to import materials from other countries.

**Fifth part: Suggestions to improve industrialized building system in Malaysia**

Do you have any suggestions to resolve the constraints of industrialized building system in Malaysia?

1- Review the concept of industrialized building system in Malaysian construction sector.

2- Improve the handbooks and the guidelines of modular coordination.

3- Improve the academic curricula in architectural and civil engineering studies concerning industrialized building system and modular coordination.

4- The role of the government is very important to enforce the use of industrialized building system in both private and public sectors.

### 3- The Interview with Dr. Surat, architectural department, Universiti Kebangsaan Malaysia (UKM).

**First part: Respondent's information**

- Name: Dr. Mastor Surat

-Specialization: lecturer& designer (architect)

**Second part: IBS Information**

Experience in industry since 1988

Type of IBS components used in projects? Pre-cast Concrete

Types of IBS projects: Social projects (schools)

**Third part: The advantages of industrialized building system in Malaysia**

In your point of view, what the advantages of industrialized building system that Malaysian construction has achieved?

1- The first advantage of industrialized building system is reducing the total construction time. For example: a school consists of 4 stories using industrialized building system completed in 8 months. The same project needs 1 year and 6 months as a normal time when using conventional method. Thus, industrialized building system can decrease construction time to 50%.

2- By using industrialized building system, the quality of buildings can be enhanced.

3- It is saver than conventional method. The accidents are very seldom

4- Industrialized building system can give very clean and tidy construction site because of manufacturing in the factories policy.

**Forth part: The constraints of industrialized building system in Malaysia**

In your point of view, what do the constraints of industrialized building system that Malaysian construction still suffer from?

The most critical constraint of industrialized building system in Malaysia is the high cost of this type of building. IBS cost in Malaysia is not effective. We need to increase the properties of cement from grade 30 to grade 50 of IBS components to increase the elasticity. IBS components need to be stronger and stiffer to be able to afford transportation process, lifting, storage and erection.

This increment in cement gives very good properties but at the same time it increases the cost of the components.

Moreover, transportation process needs an additional cost. However, if we manage to fabricate IBS components at site when we have big projects like residential and commercial complex then, we can set up the factories at site and avoid transportation cost.

**Fifth part: Suggestions to improve industrialized building system in Malaysia**

- Do you have any suggestions to improve the performance of industrialized building system in Malaysia?

1- The government have to increase the awareness of industrialized building system and modular coordination in Malaysian construction sector.

2- The educational programs in the universities have to be improved.

**4- The Interview with Ir. Noraini Bahri, General Manager, Construction Industry Development Board (CIDB).**

**First part: Respondent's information**

Name: Ir. Noraini Bahri

Specialization: civil engineer,

**Second part: IBS Information**

Experience in IBS: 20 years in IBS management

**Third part: The advantages of industrialized building system in Malaysia**

In your point of view, what the advantages of industrialized building system that Malaysian construction has achieved?

1- Time reduction is the most beneficial achievement of industrialized building system in Malaysia. This system needs shorter time because every component is produced in factory while site works can be done at the same time. CIDB have executed three houses by using industrialized building system which completed in 1 and a half months which need normally 9 months if conventional method was used.

2- This system can give very clean construction site and reduce waste material.

3- The number of workers can also be reduced, especially the foreign workers. Some of the foreign workers inter illegally to Malaysia. They did not have enough experience in construction works in their countries. When they come to Malaysia, they involve in construction sector. As a result, the quality of construction production can be affected. These problems can be avoided by using industrialized building system. The workers must be skilled and semi-skilled, and the components are prefabricated in factories, where there is a quality control, then the components are sent to construction site and installed using specific machineries.

4- Industrialized building system can provide safe construction site due to using skilled and trained workers, however, if any accident happened, the victims will be very limited because of small number of workers who work at construction site.

5- Despite all these advantages, people think that the cost of industrialized building system is slightly higher than conventional method. The truth is that the overall cost of industrialized building system is cheaper. In fact, the frameworks that are used in industrialized building system work are expensive, but they can be used for several times and cover their cost. Moreover, there is an ability to re- duce the waste materials and minimize the cost of transferring waste materials duo quality control. In addition, the exemption of the construction levy imposed by CIDB can be given to housing developers in private sector who use IBS components more than 70%.

**Forth part: The constraints of industrialized building system in Malaysia**

In your point of view, what do the constraints of industrialized building system that Malaysian construction still suffer from?

There are some problems that happen during erection process due to using closed system.

1- During the construction of CIDB's three houses, the walls, slabs, columns and beams where from different companies. The challenge was how to install theses different elements that have different types of joints with each other. The consultants of each company dis- cussed this problem and compromised it. Finally, they were able to solve this problem, and the elements were ready for installation. The standard joints are now under some researches to adapt this type of joints in Malaysian construction industry.

2- Last time, there was also leakage problem. When the use of industrialized building system started at sixties, there were leakage problems because the systems that used were not suitable for Malaysian environment. The industrialized building systems that were used in the two pilot projects were European systems. The Europeans don't use water in their toilets. On the contrary, Malaysians used to use water in their toilets and bathrooms due to different cultures. Therefore, the leakage happened in these two pilot projects. This problem has been solved and it is no longer available due to the development and the researches of our construction industry in this field.

3- Modular coordination is not applied in Malaysian construction industry. When there are standard dimensions, it will be easier for the designers, manufacturers, contactors and users to deal with industrialized building system. CIDB's designers use modular coordination checker (MC-checker) in CAD drawings to check which areas doesn't match the modular coordination system.

4- Lack of experience: the study at universities does not go deep in industrialized building system and modular coordination. It is not enough, and the students need some training. Moreover, the current architects, engineers and contractors need to be trained. CIDB offers training courses for all these specializations. In addition, the fees of training are subsidized by CIDB and become very cheap. By offering these training courses, CIDB aims to give knowledge for all these specializations, and there is no revenue gotten.

5- Payment problem. In conventional method, it is used to pay according to the executed work at construction site. However, this method is not suitable for industrialized building system as there are other needs that must be paid like setting up the factories, employing skilled and semi-skilled workers, providing machinery and moulds, and the expenditures of transportation process.

6- Some architects claimed that using industrialized building system and modular coordination may restrict their architectural creativity, but the fact said that there are many creative projects in the world have been designed using industrialized building system and modular coordination. If the architects believe in industrialized building system and modular coordination, they can design creative projects using these systems.

7- Transportation problems: The big size of IBS components and the need to specific cranes, lifters, trucks and other machineries to handle these components may add an additional cost to the total cost of the projects. As a result, this will decrease the acceptance of industrialized building system.

8- Production process: When CIDB enforced using at least 70% of industrialized building system in government projects in October 2008, most of the factories increased their capacity and productivity as there are many government projects in Kuala Lumpur and Selangor.

9- Materials and machineries: Normally, we use our local materials, therefore, there are no additional costs to import materials except some kinds of cement that we need to import. Moreover, some machineries are assembled in Malaysia and also there is no additional cost needed to import the machineries.

**Fifth part: Suggestions to improve industrialized building system in Malaysia**

- Do you have any suggestions to improve industrialized building system performance in Malaysia?

1- Legislate the regulations to enforce IBS suppliers to deliver IBS components on need by applying just-in-time principle.

2- Payment method that they normally use to follow in conventional building system is not suitable for IBS projects. The empowered persons in CIDB focused on developing a new payment form for IBS projects to bind the clients to pay at exact time.

3- Improving the application of open building system. The system that is used in Malaysia is closed system which have its limitations of not matching with the elements of other companies. Open building system provide using the prefabricated elements from different manufacturers at varied locations without any mismatching problems.

4- Applying quality supervision from the government in all IBS factories to ensure that the quality of IBS components match the standardization.

# References

Abdullah,N. G. (2009). Architectural department, Universiti Kebangsaan Malaysia (UKM), Malaysia. Personal communication, November 2009.

Agus, M. R. (1997) Urban development and housing policy in Malaysia. *Int. J. Housing Sci. Application*, 21(2), pp. 97-106.

Ali, A. A. A. (2009). Department of Civil Engineering, Universiti Putra Malaysia (UPM), Malaysia. Personal communication, June 2009.

Badaruzzaman, W. H. W. (2008). Civil Engineering Department, Universiti Kebangsaan Malaysia (UKM), Malaysia. Personal communication, November 2008.

Badawi, A. A. (2004) *The 2005 Budget Speech*. Ministry of Finance Malaysia.

Badawi, A. A. (2005) *The 2006 Budget Speech*. Ministry of Finance Malaysia.

Badawi, A. A. (2008) *The 2009 Budget Speech*. Ministry of Finance Malaysia.

Badir, Y. F., Kadir, M. R. A. & Hashim, A. H. (2002) Industrialised Building Systems Construction in Malaysia. *Journal of Architectural Engineering*, 8 (1), pp. 19-23.

Bahri, N. (2009) IBS Centre, CIDB Malaysia, Kuala Lumpur. Personal communication, January 2009.

Banico, J. E. (2009). Sales and Marketing, Cycle world Corporation SDN BHD., Malaysia. Personal communication, September 2009.

Basri, N. I. (2008) *Critical Success Factors for IBS Adoption in Malaysian Construction Industry*. Master thesis, Universiti Teknologi Malaysia.

Bensonwood (2003) *About Open Building*. Bensonwood homes.

Bernama (2006). Malaysian National News Agency. More Than Five Million Foreign Workers in Malaysia by 2010. Available at: http://www.bernama.com (Accessed: 12 August 2009).

Bing, L., Kwong, Y. W. & Hao, K. J. (2001) Seismic Behaviour of Connection Between Precast Concrete Beams. *CSE Research Bulletin*, No.14.

Chew, Y.L., and Michael (2001). Construction Technology for Tall Building. 2nd Edition. Singapore: Singapore University Press and World Scientific Publishing Co. Pte. Ltd.

CIDB (2003a) *Survey on the Usage of Industrialized Building Systems (IBS) in Malaysian Construction Industry*. Construction Industry Development Board (CIDB) Malaysia, Kuala Lumpur.

CIDB (2003b) *Industrialized Building System IBS Roadmap 2003-2010*. Construction Industry Development Board (CIDB) Malaysia, Kuala Lumpur.

CIDB (2005a) *Precast Concrete Construction*. IBS Digest. Construction Industry Development Board (CIDB) Malaysia, Kuala Lumpur.

CIDB (2005b) *Survey on the Malaysian Architects' Experience in IBS Construction.* Construction Industry Development Board (CIDB) Malaysia, Kuala Lumpur.

CIDB (2005c) *Serdang Hospital – IBS Hybrid of Precast Concrete and Steel Structures.* IBS Digest, April – June 2005. Construction Industry Development Board (CIDB) Malaysia, Kuala Lumpur.

CIDB (2005c) *Introduction to The IBS Content Scoring System (IBS SCORE) Manual.* IBS Digest. Construction Industry Development Board (CIDB) Malaysia, Kuala Lumpur.

CIDB (2007) *Construction Industry Master Plan (CIMP 2006-2015).* Construction Industry Development Board (CIDB) Malaysia, Kuala Lumpur.

CIDB (2008) *Cost Comparison between Load-Bearing IBS Blocks with Conventional Frames and Cement Sand Bricks/Clay Bricks.* IBS Digest. Construction Industry Development Board (CIDB) Malaysia.

CIDB (2009a) *Manual for IBS Content Scoring System (IBS SCORE). Construction* Industry Development Board (CIDB) Malaysia, Kuala Lumpur.

CIDB Malaysia official Portal (2009b) *What is Modular Coordination.* Available at: http://www.cidb.gov.my (Accessed: 21 December 2009).

Din, H. (1994) Industrialised Building and its Application in Malaysia, *Journal of Ministry of Housing and Local Government, Malaysia,* 1, pp. 5-10

Ghani, M. K., Hamid, Z. A., Zain, M. Z. M., Rahim, A. H. A., Kamar, K. A. M. & Rahman, M. A. A. (2007) Safety in Malaysian Construction: The Challenges and Initiatives. *Construction Research Institute of Malaysia (CREAM) - CIDB Malaysia*

Ghazali, Z. A. (2006) Modular Design Rules: part one. *IBS Digest.* Construction Industry Development Board (CIDB) Malaysia. Issue 2, pp 23- 24.

Gibbons, J. H. (1986) *Technology, Trade and the US Residential Construction Industry.* United States of America, Congress of the U.S. Special Report.

Glass, J. (1999) *The Future for Precast Concrete in Low-Rise Housing.* Precast Housing Feasibility Study Group, United Kingdom.

Habraken, N. J. (2003) Open Building as a condition for industrial construction. *20th International Symposium on Automation and Robotics in Construction ISARC 2003*, pp 37- 42.

Hamid, Z. A., Kamar, K. A. M., Ghani, M. K. & Rahim, A. H. A. (2007) The Essential Characteristics of Open Building System (OBS). *Construction Research Institute of Malaysia (CREAM), Kuala Lumpur.*

Hamid, Z. A., Kamar, K. A. M., Zain, M. Z. M., Ghani, M. K. & Rahim, A. H. A. (2008) Industrialized Building System (IBS) in Malaysia: the current state and R&D initiatives. *Construction Research Institute of Malaysia (CREAM).*

Haron, N. A., Hassim, S. & Kadir, M. R. A. (2005a) Building Cost Comparison Between Conventional and Fully Prefabricated System in Malaysia: A Case Study of Single and Double Story House. *The Professional Journal of the Institute of Surveyors, Malaysia.*

Haron, N. A., Hassim, S., Kadir, M. R. A. & Jaafar, M. S. (2005b) Building Cost Comparison Between Conventional and Formwork System. *Jurnal Teknologi, Universiti Teknologi Malaysia,* 43(B) Dis. 2005, pp. 1-11.

Haron, N. A., Hassim, S. & Kadir, M. R. A. (2005c) Building Cost Comparison between Conventional and Composite Construction System in Malaysia: a Case of Single Story House. *The Professional Journal of the Institute of Surveyors, Malaysia.*

Hashim, M. S. (1998) The Industrialised Construction System - Pengalaman (The Experience of) PKNS Engineering & Construction Berhad (PECB). *Colloquium on Industrialised Construction System.* CIDB.

Hassim, S., Sazalli, S. A. A. H. & Jaafar, M. S. (2008) Identification of Sources of Risk in IBS Project. *European Journal of Social Sciences,* 6 (3), pp. 315-324.

Hassim, S., Jaafar, M. S. & Sazalli, S. A. A. H. (2009) The Contractor Perception Towers Industrialised Building System Risk in Construction Projects in Malaysia. *American Journal of Applied Sciences,* 6 (5), pp. 937-942.

Hong, O. C. (2006a) *Analysis of IBS for school complex.* Master thesis, Universiti Teknologi Malaysia

IBS Center (2012). About IBS. Available at: http://www.ibscentre.com.my (Accessed: 4 January 2012).

Ibrahim, O. (2003) Challenges in the development of steel construction in Malaysia. *International Conference on Industrialized Building Systems (IBS 2003) Global Trends in Research, Development and Construction.* Kuala Lumpur, pp 2-8.

Ibrahim, R. (2009). Faculty of Design and Architecture, Universiti Putra Malaysia (UPM), Malaysia. Personal communication, March 2009.

Idrus, A., Hui, N.F.K. & Utomo, C. (2008). Perception of Industrialized Building System (IBS) Within the Malaysian Market. *International Conference on Construction and Building Technology (ICCBT 2008)* - B - (07) – pp75-92.

Ismail, Z. (2007) The First Industrialized Building System (IBS) Timber House (TH) in Malaysia. *IBS Digest.* Construction Industry Development Board Malaysia (CIDB), Kuala Lumpur.

Ismail, Z. & Ahmad, A. S. (2006) Modularity Concept in Traditional Malay House (TMH) in Malaysia. *International conference on construction industry.*

Jaafar, M. S. (2009). Department of Civil Engineering, Universiti Putra Malaysia (UPM), Malaysia. Personal communication, June 2009.

L. Jaillon & C. Poon. Sustainable construction aspects of using prefabrication in dense urban environment: a Hong Kong case study. Construction Management and Economics, vol. 26, pp. 953-966, 2008.

Jaillona, L. & Poon, C. S. (2009) The evolution of prefabricated residential building systems in Hong Kong: A review of the public and the private sector. *Automation in Construction,* 18 (3), pp. 239-248.

Jumaat, M. Z. (2009). Department of Civil Engineering, Faculty of Engineering, University of Malaya (UM). Personal communication, August 2009.

Kadir, M. R. A., Sulaiman, M. Y. & Ismail, S. (2005a) The Danish Trail for IBS Implementation. *IBS Digest*. Construction Industry Development Board (CIDB) Malaysia, Kuala Lumpur.

Kadir, M. R. A., Lee, W. P., Jaafar, M. P., Sapuan, S. M. & Ali, A. A. A. (2005b) Performance Comparison Between Structural Element of Building System in Malaysia. *American Journal of Applied Science*, 2 (5), pp. 1014-1024.

Kadir, M. R. A., Lee, W. P., Jaafar, M. S., Sapuan, S. M. & Ali, A. A. A. (2006) Construction Performance Comparison Between Conventional and Industrialized Building Systems in Malaysia. *Structural Survey*, 24 (5), pp. 412-424.

Kamar, K. A. M. & Hamid, Z. A. (2008) Utilization of IBS Waste Material for the Production of Concrete Pedestrian Block (CPB). *Construction Research Institute of Malaysia (CREAM)*.

Kamar, K. A. M., Hamid, Z. A. & Alshawi, M. (2009a) The Critical Success Factors for Industrialized Building System (IBS) Contractors. *Malaysian IBS International Exhibition (MIIE 2009)*.

Kamar, K. A. M., Alshawi, M. & Hamid, Z. (2009b) Industrialized Building System: The Critical Success Factors. *BuHu 9th International Postgraduate Research Conference (IPGRC)*. Salford, United Kingdom.

Kamar, K. A. M., Hamid, Z. A., Sani, S. F. A., Ghani, M. K., Zin, M. Z. M., Rahim, A. H. & Karim, A. Z. A. (2010) The Critical Success Factors (CSFs) for the Implementation of Industrialised Building System (IBS) in Malaysia. *Construction Research Institute of Malaysia (CREAM)*.

Kaur, S. (2009) Malaysian builders still far from meeting IBS targets: Board. *The New Straits Times Press*. Kuala Lumpur, Malaysia.

Kishlan, R. (2009). Baktian SDN. BHD., Malaysia. Personal communication, August 2009.

Mahmud, H. (2009). Department of Civil Engineering, Faculty of Engineering, Universiti Malay (UM), Malaysia. Personal communication, August 2009.

Masod, W. M. S. (2005) *Simulation of Allocation Activities of Logistic for Semi Precast Concrete Construction: Case Study*. Master thesis, Universiti Technologi Malaysia.

Matveev, A. V. (2002) The advantages of employing quantitative and qualitative methods in intercultural research: practical implications from the study of the perceptions of intercultural communication competence by American and Russian managers. Department of Business, City University of New York, College of Staten Island.

Mian, A. T. E. (2006) *Industrialized Building System Formation Scheduling for Public Building*. Master thesis, Universiti Teknologi Malaysia.

Migration News (1995) *Malaysia Issues Guidelines on Foreign Workers*: 2(12). Available at: http://migration.ucdavis.edu (Accessed: 4 January 2010).

Migration News (2011) *Malaysia Issues Guidelines on Foreign Workers*: 18 (1). Available at: http://migration.ucdavis.edu (Accessed: 4 January 2011).

Nagahama, M. (2000) *Japan's Prefabricated Housing Construction Industry - A Review*. GAIN Report.

Nawi, M. N. M., Nifa, F. A. A., Abdullah, S. & Yasin, F. M. (2007) A Preliminary Survey of The Application of Industrialized Building System (IBS) in Kedah and Perlis Malaysian Construction Industry. *Conference on Sustainable Building South East Asia*, November, Malaysia.

Oleiwi, M. Q. (2011) *Assessment of industrialized building system construction in Malaysia*. Master thesis, Universiti Tenaga National (UNITEN).

Pheng, L. S. & Chuan, C. J. (2001) Just-in-Time Management of Precast Concrete Components. *Journal of Construction Engineering and Management*, 127 (6), pp. 494-501.

Rahman, A. B. A. & Omar, W. (2006) Issues and Challenges in the Implementation of IBS in Malaysia. The 6th Asia-Pacific Structural Engineering and Construction Conference (ASPEC 2006). Kuala Lumpur, Malaysia.

Roztocki, N. (2001) Using Internet-Based Surveys for Academic Research: Opportunities and Problems. The 2001 American Society for Engineering Management (ASEM) National Conference, 2001, pp. 290-295

Roztocki, N., & Lahri, N. A. (2003). Is the applicability of Web-based surveys for academic research limited to the fields of information technology? In Proceedings of the 36th Hawaii International Conference on System Sciences (HICSS'03), pp. 1-8. Los Alamitos, CA: Computer Society Press.

Sarja, A. (2005) *Open and Industrialized Building*, E & FN Spon.

Shaari, S. N. (2006) IBS Roadmap 2003-2010: The Progress and Challenges. *Board of Engineers, Malaysia*., The Ingenieur Sept-Nov 2006 Issue.

Sharma, B. S. (2004) *Precast Large Panel (PLP) Construction: Some Experience from Thailand*. Civil Computing Computer Applications in Civil Engineering, Thailand.

Tahir, M. M., (2009). Architectural department, Universiti Kebangsaan Malaysia (UKM), Malaysia. Personal communication, August 2009.

Tat, C. W. & Hao, H. (1999) Precast planning for Singapore. *Prefabrication Technology Center 4th Anniversary Seminar Singapore*.

Thanoon, W. A., Peng, L. W., Kadir, M. R. A., Jaafar, M. S. & Salit, M. S. (2003) The Experiences of Malaysia and Other Countries in Industrialized Building System in Malaysia. *International Conference on Industrialized Building Systems*. Kuala Lumpur, Malaysia.

Trikha, D. N. (1999) Industrialized Building System: Prospect in Malaysia. *Proceeding of World Engineering Congress*. Kuala Lumpur.

Trikha, D. N. & Ali, A. A. A. (2004) *Industrialized Building System*, UPM.

Warszawski, A. (1999) *Industrialized and Automated Building Systems*, E&FN Spon.

Wee, E. C. (2006) *Application of Acotec Industrialized Building System in Malaysia's Construction Industry.* Master thesis, Universiti Teknologi Malaysia.

Wise Geek (2009) *What is Just-in-Time Manufacturing.* Available at: http://www.wise-geek.com (Accessed: 5 October 2009).

# Glossary

IBS     Industrialized Building System

CIDB Construction Industry Development Board

CMU Concrete Masonry Units

HDB Housing and Development Board

PTC Prefabrication Technology Centre

NHA National Housing Authority

PLP     Precast Large Panel

ACA     Accelerated Capital Allowance

MC      Modular Coordination

MS      Malaysian Standard

Acotec Advanced Construction Technology

CSFs Critical Success Factors

PKNS Perbadanan Kemajuan Negeri Selangor

CREAM          Construction Research Institute of Malaysia

SIRIM Standards and Industrial Research Institute of Malaysia

JKR     Jabatan Kerja Raya

1M      Unit or basic module for Modular coordination that equals to 100 mm

NNI     Nederland's Normalisatie Institute

UBBL Uniformed Building By Laws

CIS     Construction Industry Standards

R&D     Research and Development

BIPC Building Industry Presidential Council

ABM Akademi Binaan Malaysia

USA     United States of America

SPSS Statistical Package for Social Science

www.ingramcontent.com/pod-product-compliance
Lightning Source LLC
Chambersburg PA
CBHW030845180526
45163CB00004B/1454